"As it states in this wonderful book, 'Music learning is the gift that keeps on giving.' It's true! I've not only witnessed it, I've experienced it since birth. But I've never seen it laid out so simply and thoroughly. All kids benefit from music. *The Music Advantage* not only shows you how, it shows you why. This book is important for our future."

—Victor L. Wooten, author of *The Music Lesson*

"*The Music Advantage* is the book I always wanted to exist, and now it does! It reveals new insights, showing the power and magic that music has on the neurological well-being and development of children, and all of us."

—Stephon Alexander, PhD, author of *The Jazz of Physics* and Professor of Physics, Brown University

"Dr. Anita Collins's insights, based on extensive practical research, highlight the irreplaceable role music plays in the full education of a child. *The Music Advantage* is essential reading for parents, educators, and policy designers."

—Benjamin Northey, Principal Conductor in Residence of the Melbourne Symphony Orchestra

"*The Music Advantage* is fascinating. This book helps parents to understand how valuable learning music is to a child's development, starting in preschool up until their first job interview."

—Allie Ticktin, MA, OTD, OTR/L; founder of Play 2 Progress

"Dr. Collins's exuberance and enthusiasm shine through as she persuasively argues for the importance of music in every child's education."

—Alan R. Harvey, author of *Music, Evolution, and the Harmony of Souls*

"Drawing from her dialogue with scientists, music teachers, and her own relationship to music, Anita Collins makes an argument for music education told through stories we can see ourselves in."

—Nina Kraus, PhD, Hugh Knowles Professor of Communication Sciences, Neurobiology, and Otolaryngology at Northwestern University

"Two decades of neuroscientific and psychological research into the effects of learning music on our brains are condensed into a highly readable, fascinating, and important book. Important because the overwhelming evidence is that music education will benefit every child—and thus it should be part of their overall education." —*Loudmouth* magazine

"Shows the value of music education on brain development and includes practical tips for harnessing melody and rhythm to increase learning in all areas." —*Library Journal*

"A grand volume that avails us of yet another level of groundbreaking research highlighting the extraordinary benefits music offers to the positive growth and development of *every* child. Dr. Collins, you have proven once and for all that one person makes a difference. Thank you."

—Tim Lautzenheiser, Vice President of Education, Conn-Selmer, Inc.

THE
MUSIC
ADVANTAGE

How Music Helps Your Child
Develop, Learn, and Thrive

DR. ANITA COLLINS

A TarcherPerigee Book

tarcherperigee

an imprint of Penguin Random House LLC
penguinrandomhouse.com

First TarcherPerigee trade paperback edition 2022
First published in Australia in 2020 by Allen & Unwin
Copyright © 2021 by Dr. Anita Collins
Penguin supports copyright. Copyright fuels creativity, encourages diverse voices,
promotes free speech, and creates a vibrant culture. Thank you for buying an authorized
edition of this book and for complying with copyright laws by not reproducing, scanning,
or distributing any part of it in any form without permission. You are supporting writers
and allowing Penguin to continue to publish books for every reader.

TarcherPerigee with tp colophon is a registered trademark of Penguin Random House LLC.

Most TarcherPerigee books are available at special quantity discounts for
bulk purchase for sales promotions, premiums, fund-raising, and educational needs.
Special books or book excerpts also can be created to fit specific needs.
For details, write: SpecialMarkets@penguinrandomhouse.com.

The Library of Congress has catalogued the hardcover edition as follows:

Names: Collins, Anita (Anita Marie), author.
Title: The music advantage: how music helps your child develop, learn, and thrive / Anita Collins.
Description: New York: TarcherPerigee; Sydney: Allen & Unwin, 2021. | Includes index.
Identifiers: LCCN 2020041086 (print) | LCCN 2020041087 (ebook) |
ISBN 9780593332122 (hardcover) | ISBN 9780593332139 (ebook)
Subjects: LCSH: Music—Psychological aspects. | Child development. |
Music—Instruction and study—Psychological aspects. | Musical ability in children.
Classification: LCC ML3838 .C648 2021 (print) | LCC ML3838 (ebook) | DDC 372.87—dc23
LC record available at https://lccn.loc.gov/2020041086
LC ebook record available at https://lccn.loc.gov/2020041087

ISBN (paperback) 9780593421451

Printed in the United States of America
1st Printing

For Michael and Ellie

CONTENTS

PART FOUR: Almost a grown-up, 12–14 years to adult

INTRODUCTION

At the end of every academic school year there is some type of celebration: a speech night, presentation assembly, or awards ceremony. Whatever it was called at your school or at your own child's school, it probably involved music. It might be a live performance by students or just recorded music selected for the event, but you can bet that at some point during the proceedings there will be music.

As a music teacher, I always marvel at the polar opposite experience of my colleagues and friends in the math, physical education, and English departments around presentation night. They appear to be gently winding down, reports finished and end-of-year decisions made. Meanwhile, music teachers like me are rapidly revving up to the most-anticipated public, and possibly judged, performance of the year. The entire school community, ready and waiting to be impressed and entertained.

Imagine a mass of musicians from the ages of ten to eighteen in an orchestra in front of the stage, and then me and my other music magic makers sitting facing them. With a running sheet many pages long and every minute of the event scheduled, we are tightly wound and ready for action. After the national anthem is played and the speeches are done, I always breathe a big sigh of relief. The show is rolling and there is a lull as the first awards are presented. If I am honest, it also gets a bit boring, so to keep the adrenaline up I play a game: How many musicians got prizes for subjects other than music this year?

For many years running the answer has been close to the same. Seventh grade is announced and ten students are awarded certificates for high achievement in science or English, while some of them receive awards for consistent achievement across all of their subjects. I would get my scorecard out, usually the back of the program, and keep a tally as they walked across the stage. Musician, musician, musician, musician, played trombone for a while but then gave up so that counts, nonmusician, musician. Do you want to guess how many on average out of ten I would get in each grade? I present this question to music educators in lectures all the time and they are usually spot-on—seven or eight of the top ten prizewinners either had or were continuing a significant involvement in music.

At that point I could just congratulate myself and my music colleagues on helping to produce the best and brightest in the school, but I often asked myself a different question: Did learning music help these students to achieve at the highest levels in their academic work, or were these students already smart to begin with and therefore drawn to learning music? Was it the chicken or the egg, nature or nurture?

Little did I know that these end-of-year musings would take me, a music teacher who had very little interest in science in school and struggled with reading until I was nine years old, into the world of neuroscience and psychology. I have traveled the world looking for answers to these questions, visiting labs working in the field of neuromusical research to understand how music learning impacts on brain development. While I am a teacher of high school–aged students, I have been amazed at how music listening and music learning at all ages, from day one of life to our final day, has the ability to grow, change, and repair our brains.

What I have found out is far more complex than just the chicken/egg or nature/nurture answer. I have learned how babies hear their mother's voice as if it was music, how the necessary neural connections for reading are active when a toddler can keep a steady beat on a drum and how music can remold the brain after injury or trauma. Turns out that the teaching of music—the chord progressions and performance anxiety and ability to practice effectively—isn't just content and skills. These parts, in fact, contribute to growing a human, and suddenly my role in each one of my students' lives became so much more important than I ever knew.

But there was no point in just me knowing about this research. It needed to be shared, but it seemed to me that there were a few big hurdles in the way.

First, research is most commonly shared via peer-reviewed articles in academic journals. These articles are often long, hard to read, and full of jargon, and the general public would struggle to find them easily.

Second, neuroscientists and psychologists are primarily using music listening and learning to understand how the brain develops and learns, not to understand how to teach music in schools or how to transfer the findings to parents and their children. There is a gaping hole between what is published and what is happening in the world outside the lab.

Third, scientists don't say the word "prove." As part of the scientific method, the findings of an experiment will only ever suggest, highlight, point to, or explore; one study will rarely if ever prove something definitively. And that's the way it should be. We need to have multiple researchers testing multiple theories in multiple ways multiple times to really understand any phenomenon. Another hurdle is that this research is based on human beings who are all unique and a product of their genetics, personality, and experiences. To identify how music and music learning may impact on brain development, researchers need to work with averages and group data, yet every child learns slightly differently and has a different mixture of predispositions and life experiences that have shaped them. For every finding that is reported, there will be an outlier whose experience goes against the norm.

Fourth, research is often incredibly detailed and difficult, and that is why researchers can spend their entire lives pursuing, in one way or another, the complex answer to a simple question. The tricky part in trying to share detailed, expert research with someone who isn't trained in the field, hasn't done research, and is trying to apply it to their own experience is the ever-present and pretty much unavoidable danger of oversimplifying.

This book is my attempt, in the face of all of these hurdles, to open the door for you, the everyday reader, parent, teacher, and student, to the world of neuromusical research and explain how its findings will in some cases reinforce and in other cases challenge what we think we know about music and music learning. It is just a start, a crack in the door, a shaft of light maybe, but what lies behind the door is a world I have certainly found fascinating, and I hope you will, too.

Useful stuff to know before you start

First, some definitions for the purposes of this book. *Music listening* is just what it sounds like, and our engagement with music can be either *passive* (background music) or *active* (music we put on in the car and sing along to). *Music education* is the more formal learning of music, often through a musical instrument. This might mean any of a combination of weekly lessons, learning in a musical ensemble, reading music, and performing throughout the school year and over multiple years. You will see as you read this book that I have opted for the term *music learning*, and the reason for this is to get us thinking about the many ways music education happens. *Music experience* can have different meanings at different ages. For young children it looks like play, banging pots and pans with a wooden spoon on the kitchen floor or teacher-led activities that look from the outside like games, and this type of exploration morphs seamlessly into more formal music learning. Music experience for older children includes one-off or short-term learning, such as a visiting ensemble to the school or a four-week drumming immersion program. Music experiences augment music education/learning but are not a substitute for them when it comes to the impacts on cognitive development.

Next, some pocket history. The neuromusical research field started in its current form in the mid 1990s when new technology (fMRI, PET, EEG, and so on) allowed researchers to watch the human brain function in real time. It began with getting participants to listen to music and then watching how the brain processed sound. In a TED-Ed video I wrote in 2014 we called this "fireworks." I wish I could own the description but I actually heard it from a researcher explaining that some neuroscientists found out by accident that music processing set off more activity in the brain than anything else they had seen at that point. This happy accident meant that researchers started to use music as a tool to understand how the brain worked.

Then researchers noticed that the brains of musically trained adults seemed to both look different and function differently, and indeed more effectively, than nonmusically trained participants. So they began to do research with two groups undertaking the same task, one being adult musicians and the other being nonmusicians. Now, remember that while the researchers were using musically and nonmusically trained participants as comparison groups, they were actually trying to understand anything from neuroplasticity to auditory processing to working memory to attention span ... the list goes on and on. One of the earliest findings was that the musically trained participants used less brain activity to complete a task and often executed it with greater accuracy. Did music learning train the brain to work more effectively and efficiently or did people with brains that were already effective and efficient find learning music fulfilling and enjoyable?

The next step was to investigate how music learning might impact on children's brain development. Did it matter when children started to learn music? Was there a minimum amount of time required for music learning to affect cognitive development? Was one method or pedagogy better than another and did the instrument choice matter? What were the mechanisms in music learning that changed the brain, or conversely, what were the predisposed mechanisms such as genetics and personality factors that made learning music easier or more affecting? Again, the studies tended to compare musicians with nonmusicians, and one factor that may have had an impact on the findings was that sometimes the musician groups were self-selecting, meaning they were made up of children (and probably their parents) who chose to take music classes.

This period of research still found some amazing things, and the discovery that our auditory processing system is the largest information-gathering sense in our brains was a revelation. Our sound environments are crowded not just with the volume of sounds but the variety, too. Even the idea that having the television on at home at a low level could impede a toddler's speech acquisition due to auditory overload is a new understanding that we would not have had if researchers hadn't studied the effect of music on the developing brain.

In the mid 2010s we started seeing the gold standard of scientific design: the randomized longitudinal study. In these studies children were randomly assigned to music learning, to another activity such as sport, and to no activity at all; then they were followed and tested by the researchers for three to five years. Still more amazing revelations and understanding came to the fore, but again the purpose was not necessarily to prove or disprove the positive impact of music learning on brain development but to illuminate the phenomenon in light of what we know about the brain and how we view music listening and music learning in the lives of our children—and, increasingly importantly, in the lives of adults.

The research isn't over yet. Actually, I think it will probably never be over, much like a realization I had in the final stages of my PhD: you never finish a PhD, you just stop working on it. In 2018 and 2019, several robust research reviews were undertaken looking across the findings to see what the overall impact of music learning has been on brain development, and the "consensus" from these first studies is anything from strong, mixed, to small. I love watching this field of research unfold as researchers grapple with the difficulties of human-based research and the complexity of music learning. I believe the next five years will see some extraordinary developments that will challenge and inform both the scientific and educational worlds.

Last, a word on limits. I am a researcher and a teacher and a communicator. In writing this book I feel a tension between wanting to get this research into the world of the parent, teacher, and student and not wanting to dumb down or simplify the research, where detail and definition are so important. I have also not gone into depth about the size of the impact of every study I have discussed and it is important to note that the impact of music learning on the given skill that was measured may be only small or moderate. There is still a great deal of debate and ongoing research in the neuromusical field about the size of the impact as it can vary across studies and methodologies. As you read through this book it is important to be aware of the nuances and limitations of research in general. So to my more expert readers, this is not an academic document with citations for every statement or study—it is meant for the general reader, the parent, the teacher, and the student, to fuel a desire to learn more. With that in mind, each chapter finishes with just a small list of articles and studies where readers can read more, rather than a huge reference section. There are other academic and professional publications catering to that side of things.

For my nonacademic readers, hello, this book is for you! Please read it with an open and questioning mind. As you hopefully laugh and maybe cry, say "aha" and reread something that doesn't make sense the first time around, remember to marvel at the complexity and don't reach for the sound bite. A researcher once told me that we don't even have ways of thinking yet that can comprehend how incredible and elegant the human brain actually is, so if our brains aren't simple or generic then neither should the way we think about them be.

A book of stories and studies

Each chapter begins with a story and that is because, as a teacher, I have learned the power of story to explain a concept. The stories are all real and come from my own experiences, or from experiences that have been related to me. All of the names have been changed and in several cases the specific details have also been changed to protect the individuals and schools involved. I also share stories from my travels to labs and interactions with researchers—I am so grateful for the time and energy these researchers, students, and teachers gave me.

Many of the stories come from my visits to research labs all over the world. Along with reading new research daily, often before I even get out of bed in the morning, I have also visited sixteen labs in the United States, Canada, Germany, Sweden, Norway, and Australia and interviewed more than a hundred researchers, from assistants in labs to the professors leading multimillion-dollar facilities. I myself participate in as many of the experiments as they let me, I observe children and adults in experiments and visit sites where the music programs are being conducted. This helps me to educate myself, to hear about the research that didn't make it into the academic journals, and to see what is next on the horizon.

As you read each chapter, I hope that parents and grandparents will understand what your child or grandchild is experiencing as they learn music but also grasp that when things get tough, your little one's brain is learning how resilience feels. For teachers, I hope this book will make the learning process clear in a more individualized way, especially when it comes to how your students learn through their ears. For school leaders, I hope this book will inform your understanding of the place and purpose of music learning for every one of your students. For policymakers, I hope that this research, embedded in the real-life stories of the children you create policy for, will generate a shake-up of some of the current practices and curriculums in our education systems. And finally, to the readers who are one or all of those roles mentioned above, I hope this book explains the impact that music and, if you were very lucky, music learning have had on your life so far, and may just have in the future.

THE
MUSIC
ADVANTAGE

PART ONE

FROM BIRTH TO BIG SCHOOL
0 TO 5 YEARS

1

A WINDOW TO THE WORLD

How your baby hears music in everything

"*No!*" I yelled.

Well, to be fair I didn't really yell in the traditional sense of shouting. My voice deepened, and the sound on the front of the word was very heavy as I threw my voice from the kitchen bench I was standing at to five meters in front of me, landing slap-bang on the head of my baby daughter—who had just reached for a sharp object.

Let's analyze the sound I made a bit more. We all know that sound. When I was lecturing to preservice teachers I would call it their "teacher voice" and when I was teaching in schools it was the voice I would use to stop a group of teenage boys in their tracks at fifty paces. It was the sound my father heard when he was wooing my mother, who was a young teacher at the time, and he decided to surprise her at the school fence one lunchtime when she was on duty. "How could such an incredible sound come out of such a small woman?" he would wonder as he recounted the story.

It is the sound of authority. It is the sound of warning or jeopardy. It is the sound that, no matter what language you speak, you pay attention to.

Why? And how was it that my very young daughter, with a first-time mother who hadn't quite developed that keen eye to predict how an infant could turn a pen into a weapon against themselves, understood what I had said? And not just understood it but interpreted the meaning of that single word as a signal to stop reaching for the pen and to pay heed to my warning.

In this chapter we will wrap our heads around the idea that to our babies, all sound is actually music-like and they process it in their brain using some of the fundamental characteristics of music. You don't need to become a music whiz overnight—I promise I will introduce the terms slowly and with meaning. Here is your first music lesson in a nutshell. Music is made up of ingredients, but we don't often hear all the separate ingredients, we just enjoy the finished meal (or song, or piece of music). But if you did have to take it apart, back to its individual ingredients, the more common ones are:

- *pitch*—how high and low the notes are
- *rhythm*—how long and short the sounds are
- *dynamics*—how loud or soft the sound is and how that changes

- *timbre* (or tone color)—which is the source and quality of the sound (the source of the sound might be a cat whose meow has the quality of a harsh squeal when her tail is stepped on).

Now let's put those new terms to work!

The music in my words

There were many musical elements to my "no." The *pitch* (the high and low) was lower, the *rhythm* (the length) was shorter, the *dynamics* (volume) was louder and the *timbre* (the quality of the sound) was harsher. After I took a few breaths, removed the offending item to a safe distance from my baby and went back to doing the dishes, I thought about those elements.

Maybe I was just a musical nerd (yes, really!). Here I was, a musician, a music teacher, and now a music researcher halfway through her thesis on music and the brain. It is just possible I was hearing music everywhere.

Yet the question stuck in my head. How was my daughter, or any child, understanding the meaning of language when they couldn't speak it yet? How could I on hearing someone speak in another language guess how they might be feeling or understand if they were overjoyed, amused, or distressed? How do our brains understand sound?

Understanding sound

Sound is a fascinating thing, and until recently it has been an undervalued sense. If I stand in front of a room of nonmusically trained people and ask them which of their senses gathers the most information, the majority would say their eyes. But we now know that the sense that gathers the most information, never turns off, and is active right from birth (and possibly before) is our hearing. More precisely, the auditory processing system in our brains. We know this through studying the auditory processing systems of adults right down to babies who are just hours old.

At birth, a healthy, full-term baby can't see very well, can't move with any intention or control, has taste buds that are all about milk, and her or his sense of smell hasn't had a lot of exposure to the world. But their hearing is perfect. What I mean by perfect is that it is primed and ready to start taking in information and, with the absence of the other senses pulling their weight, a baby's ears are more important than we might have given them credit for.

The big question for researchers is: How do they make sense of sound? What is sound to a baby and how does it turn into information?

The answer is that a baby's brain understands sound through its musical elements or characteristics. Let's look at this more closely. I'm going to ask you to stop reading for a moment and close your eyes. Now listen. What are the sounds around you? I can hear my air conditioner whirring, very faint car sounds from the road near my house, and every once in a while my daughter will let out a laugh in response to the TV show she is engrossed in. Now listen again and try to hear the music in the sounds.

For me the air conditioner is constant, so it is like an organ droning. The cars accelerate to reach speed so the pitch rises, the dynamics (volume) rise, and the sound becomes more complex. Then they change gear. If I suspend my rational understanding that the sound means a car is accelerating then changing gear, I can hear the melody of the car sound. Around me and you right now is music—if you let your brain, and auditory processing system, think of it that way.

Take that little moment you have just given yourself and now apply it to a baby who is only a few hours old. She is gathering information from sound by processing it into many parts simultaneously. At birth her brain already has the connectivity to process sound in multiple ways.

The three-dimensionality of sound—or imagining sound as a squishy ball

A baby's brain can process sound in terms of its parts, how those parts inform and interact with each other and in terms of its whole. Put another way, imagine looking at a picture and being able to take in the effect of the whole image at the same time as looking at the most intricate parts of the picture and how they relate to one another—things like the colors the artist has chosen, the nature of the brushstrokes, and the way they have composed the scene. This is super hard to do with our eyes but our ears manage it like it is child's play (like my pun?).

Here is just one of the aspects of sound processing that is mind-boggling. When we listen to music our brains are processing a huge amount of information both simultaneously and over time. When we hear the first note of a pop song our auditory processing network is pulling apart the sound for all of its elements or characteristics. What I mean by that is that our ears are identifying the different instrument sounds and if they are high or low. But there is more. Our auditory processing systems are also assigning a role to each part, for example, which part is carrying the tune, such as the singer, or which one is providing the rhythm, often the drum kit. But there is more. Then our auditory processing network puts all those parts back together and hears the music as a whole to determine how the parts are interacting with each other. But there is more.

What I described above is one moment, one single note, in a piece of music. Imagine this is one squishy ball. What happens next is incredible. While our auditory processing networks continue to dissect each individual moment in the music, it also analyzes the musical moment, or squishy balls, in relation to each other. How did the note in the singer's part change or stay the same? Or what happened in the drum kit part when the chorus started? These elements are all processed at the same time to extract meaning. But there is more.

As this superfast extraction and reformation processing (the auditory analysis of the music) is going on—which I always imagine looks like lots of squishy balls bouncing around at once—other parts of our brains, including our emotional, perception, and sensory networks, are also extracting information from the sound, making us feel different emotions or shifting our mental state. I imagine all the moving squishy balls take on different colors and hues. But there is more!

Then our body decides to get in on the action. All this auditory analysis and emotion shifting happens and, if we like it, our bodies respond by tapping our foot or finger or bobbing our heads to the beat. As a response our immune systems also kick in and release the dopamine we enjoy when we listen to music.

But here's the thing: you hear and process musical sound every day. Sometimes you might do it actively where you choose to engage with a piece of music and actually listen. Most of the time you will do it passively when music is playing somewhere in the background (in a shopping mall or while you wait on the phone on hold, for example). But does it feel like a basketball arena full of really complex cognitive work? The reality is that you probably don't even register that it is happening.

Sound is food for the brain

The big concept to wrap our heads around is that for tiny little babies, sound is food for their brains. Sound is a cognitive nutrient (other nutrients include touch, facial expression, and eye contact) that helps babies begin to understand their world. The nutrients are processed in much the same way as food is broken down by our bodies into substances like proteins, carbohydrates, and sugars. For the auditory processing networks in babies' brains, cognitive information like pitch, rhythm, and timbre are the nutrients to help them thrive.

From this cognitive information, babies learn a lot of things. One of the first is who their carers are. Every person's voice has a unique musical signature, a combination of pitch, rhythm, tone, and volume that is individual. Before the display of a mobile phone told us who was calling, one short "hi" would often let us know who it was. Even with caller ID we can still generally tell what kind of mood the other person is in just by the tone of that first word or two (try it next time you hear your parent, partner, or child on the other end of the phone). Our brains are processing the unique musical signature of voices.

Babies have to be able to do that right from the beginning of life. Helpless and needing to be cared for, they put all their trust in a key caregiver and identify who that person is by registering the musical signature of that person's voice.

They also need to not just identify that voice but understand the many nuances of it. Often we speak to our babies in a very specific tone, one that is full of positivity, emotion, and love. Along with facial expressions and eye contact we are using sound to help develop a bond with our child. Take a look next time a baby is in the arms of a new person he hasn't met before. He may look enthralled or like he is looking at an alien, take your pick, but what he might be doing is registering that person's musical signature to answer the simple question: "Who are you?"

Too much sound

We have established that sound is actually music inside our brains, but is it possible to have too much sound and therefore too much information?

I mentioned earlier that our hearing or auditory processing may have been our undervalued sense in the past. That has changed for a number of reasons in the last few decades. We now have the technology to not just measure what the ear hears (or doesn't hear, as we get older) but also what the brain then does with the information that we hear. This has opened up a world of understanding into how the brain is connected, how it develops, and how it changes over a lifetime. Studying music—both how we listen to it and how we make music—has been one of the tools used by researchers to grow this understanding.

At the same time, our world has become louder. The amount of sound, which is called noise in this context, in our everyday lives may be far more than our brains have evolved to process. The problem is not only the volume of the sounds but the variety of them that we need to digest every day. As I sit outside at a cafe writing right now I can hear cars passing, the sound of an engine running, and the tires on the road surface getting louder as they come closer. There are people talking behind me; I can't hear the detail of their words but two older men are having a conversation to my left and a younger woman is on the phone to my right. I know what they are doing by the melody and rhythm of their voices. I could go on describing the intermittent hammering, coffee machine gurgles, and variety of footstep rhythms, but you get the picture. My ears are hard at work processing all of this information while I think and type, open to potential dangers or changes in circumstances.

This increasing level of auditory information may be over-whelming our processing capacities and as such our brains might just be getting better at screening out certain sounds entirely. What is not known is if auditory overload leads to excessive screening out of sound, not just the sounds our ears find offensive or that overload our processing systems.

Our hearing health may be declining. So much environ-mental noise and other factors like in-ear headphones may be impacting our hearing health. You might think this isn't a problem as hearing-aid technology improves, but loss of hearing is an indicator for all manner of other cognitive issues. Think about it: If our auditory processing network is a signif-icant information gatherer for our brains and the very organ that collects that information starts to shut down, what might our brains decide to do?

All this research has helped us understand some incredible and important things about our own and our children's auditory health. One of the most important of these is understanding that music is a fundamental part of being human, that we use the built-in network in our brains to begin to understand the world from birth and that our world is full of music right from the start of life. We even use our music processing networks to help us learn how to speak and then how to read, but more on this in Chapter 8.

The sound world of a child

The next challenge is to think about the sound environments we create, allow, and control for our young children. Sound is information for a very young child's brain, so silence or very little sound is not the answer—that is just starving the brain of cognitive nutrients. The answer is to provide sound variety.

Here is a good way to start. Sometime today, take five seconds and listen to the music of your own sound environment. How high or low are the sounds (pitch), what is the rhythm of the sounds and is there repetition to the rhythm or is it very unpredictable? If you had to describe the timbre (quality), what words would you use and what is the source of the sound? For example, I often find in cafes that I get unnaturally distracted by a particular person's voice. It grates on me and I regularly move to get away from them. Often their voice isn't loud but there is something about their musical signature that makes my ears scream.

Try the above exercise to start hearing, really hearing, your own sound environment. I apologize now because I am about to open up a door to your sound environment that you won't be able to close. However, in the end I hope you will thank me because situations that may have really bothered you before for seemingly no good reason will now be explained, and you can do something with that knowledge.

When it comes to children, especially young children, auditory overload can happen very quickly. Of course, they won't be able to say to you, "Mommy, I'm in auditory overload"; they will let you know through their behavior or choices. And just as every child is an individual, every reaction to auditory overload is different.

Children put their hands over their ears not always to muffle loud sounds but often to restrict the amount of information they are asking their brains to process. I have also seen young children seek to be in the middle of a complex sound environment by crawling into it and reveling in the swirl of information. Conversely, young children can crawl away physically, or shut down mentally, to adjust the amount of sound they are processing. To add to the challenge for parents, auditory overload can happen very quickly—a little baby might love the sound of her rattle for a minute and then push it away and recoil from the sound in the next. The trick is to watch their reactions and understand that their sound environment may be affecting them more than you think.

The simplest habit we can get into is adjusting our children's sound environment when we speak to them. For our young ones to master the first hurdle with language they need to be able to separate speech from all the other noise around them. If the radio is on in the car, the TV is on as background noise, or there is a high level of other language sounds in the room when we are trying to speak to our children, we aren't helping them to gain control of their auditory processing. Sound is music, and music is information for our brains.

Further reading

Hallam, S. (2016) "Infant Musicality," *The Oxford Handbook of Music Psychology*, p. 387.

Hannon, E.E., Lévêque, Y., Nave, K.M., & Trehub, S.E. (2016) "Exaggeration of language-specific rhythms in English and French children's songs," *Frontiers in Psychology*, 7, p. 939.

Koelsch, S. (2011) "Toward a neural basis of music perception: A review and updated model," *Frontiers in Psychology*, 2, p. 110.

Norton, K. (2015) "Womb to Tomb," *Singing and Wellbeing: Ancient Wisdom, Modern Proof*, Abingdon: Routledge, p. 42.

Perani, D., Saccuman, M.C., Scifo, P., Spada, D., Andreolli, G., Rovelli, R., Baldoli, C. & Koelsch, S. (2010) "Functional specializations for music processing in the human newborn brain," *Proceedings of the National Academy of Sciences*, 107(10), pp. 4758–63.

2

THE POWER OF LULLABY

Why your baby responds to music

Me: "Do you sing to your baby?"
Father: "Yes, but only when no one can hear me."
Me: "Your baby can hear you."
Father: "But he doesn't care—he doesn't know that my singing is terrible."

This is the sort of exchange I have had with dozens of parents. What I should also include are the nonverbal signals and the prosody (the melody and rhythm) of the father's voice that also give me a great deal of information regarding how the father feels about singing to his infant son. Let me add those in.

Me: "Do you sing to your baby?" *(open face to hopefully convey genuine interest and zero judgment)*

Father: "Yes." *(big smile and a single word infused with recollection of a joyous experience, then a pause)* "But only when no one can hear me." *(flat tone, a bit of head shaking and a tiny step backward with a straightening of back)*

Me: "Your baby can hear you." *(quizzical, nonjudgmental tone but conscious to look off to the side and minimize any challenge)*

Father: "But he doesn't care." *(hands held open, slight shrug of shoulders)* "He doesn't know that my singing is terrible." *(slight laugh in voice and small return to recollection of a joyous experience)*

I can only go on my own interpretation of this exchange, but for me I see a father who has experienced a profound connection with his infant son through singing and who also may not feel completely comfortable sharing the joy that gave him with another adult, let alone a stranger. He loves it, but he's not sure if he should. If I asked a mother the same question, the answer and underlying value judgment may be very different. In Western cultures singing to your baby can be perceived as more appropriate or acceptable for a mother, especially in those first few months of life. It is often suggested as a good technique for soothing a baby to sleep, and while of course fathers use this technique, I wonder if singing in general carries less judgment for mothers than fathers.

I also wonder about the "no one" listening to him: Is it his partner, maybe his other children, or the next-door neighbor? Perhaps it's the imaginary pop star judge in a red swivel chair who says, "Love your enthusiasm, but your singing isn't so great, dude!" I have never had the guts to ask that of one of the parents I have posed the singing question to, but maybe I should.

Where did we get this idea that we had to be a good singer, as judged by another adult, to be allowed to sing to our babies in public? This chapter will look at the research that examines why, from an evolutionary perspective, we might sing to our babies.

If you have a voice you can sing

I had this experience just this morning. A fourth grade violin student arrived at a rehearsal. He had become a bit confused and didn't realize he was actually going to be in the choir for the concert.

"I play violin, I don't know how to sing," he said.

"If you can speak you can sing," I replied.

Inherent in this statement from Mr. Fourth Grade were many more questions for me. How strongly do we identify with being a singer or not? If we play an instrument, are we unable to be a singer, too? And when we decide whether we are a singer or not, what influences that decision? (For those having flashbacks to horrible choir experiences or someone saying you have a "bad" voice, I feel your pain and hang in there.)

The idea that we should hide or let professionals like the Wiggles take care of the singing in our children's lives is a relatively new one. Yes, we know there are people who have incredible voices, seemingly bestowed by a higher entity—Dame Nellie Melba, Freddie Mercury, Nina Simone, and Frank Sinatra, to name but a few. But there is a difference between a predisposition for a great voice that, with proper training, turns into an incredible voice and the kind of voices we use as parents and caregivers to connect with our young ones.

You are your baby's favorite rock star

This is a phrase I use often with parents. It started by accident when I was talking to a parent who just didn't seem to be moved by all the scientific research I was sharing with her. There was something about being her daughter's first and favorite rock star that set her free from her self-imposed judgment. Maybe there was even a little strut as she walked away from me!

From an evolutionary perspective, music and singing have a very ancient human history, at least as old as language. Babies understand the world through their ears as rhythm, pitch, contour, and timbre and they use sound to identify the important things, like who are their primary caregivers, who is part of the family or tribes and, possibly most importantly, who they can trust. One of the most effective mechanisms humans have to convey that information is through song.

Why song? Why not interpretive dance or pictures? The simple answer is that babies need to know who the people around them are as soon as possible. At birth, babies can only see vague shapes, identify faces, some movement, and strong colors. Although identifying different faces is a high priority for babies from birth, they still struggle with facial expressions for the first six months of their lives. Similarly, they can't move with great control or intention and their smell and taste libraries are only just starting to gather information. But their brains are wired for sound, and the sound of your voice is their favorite station.

Each voice has a specific auditory signature, a combination of resonances unique to you. Typical babies' brains have the capacity to process and attribute their significant people's voices to their role in their lives. This is why on meeting a new adult who speaks to them they may look a little dumbfounded. However, what's really happening is that the baby's auditory system is working very hard to answer questions like: "Who are you?" "Where do you fit into my world?" "Can I trust you?"

Now think about the types of voices we use with babies, that hyper happy tone reserved only for them. It is a tone that is infused with positivity and emotion (as I mentioned earlier) and musicality, the pitch is higher and wider, like a very jumpy melody, the dynamics are often louder, and the rhythm is longer and avoids quick sounds. This is called motherese or parentese, and it is a way of communicating that is halfway between our quite narrow and emotionally subtle language voices and a full-throated sing-along to Adele in the car (when you are on your own, of course).

The musical life of babies

Motherese and parentese have been studied extensively to help us understand how babies bond, understand language, and process auditory information. Also called infant-directed speech, it is the act of speaking to a baby in order to maintain their attention or to soothe them. The first example of parentese that pops into my head is a song I would sing to my baby in certain frustrating moments: "Your belly is full and your nappy is changed, you're wrapped up and safe, but you still want to cry, tell me why, tell me why." Catchy, and probably also a way of soothing myself.

For a baby, sound is music and music is information, and song is far more interesting than speech. Why? Because speech at this time in their lives is a group of sounds that make no sense, whereas song has a lot more variation and information because of the inclusion of rhythm and melody to the speech. Babies prefer song or parentese over speech and can maintain an interest in it for far longer. Being able to maintain attention for longer than a few seconds allows babies to take in more information and is the very basis of learning. The more practice we give babies through singing to them, the longer their attention span may become.

While babies may prefer song over speech to begin with, they are still using song to start to decode language. Learning to separate speech sounds from all the other noise around them can be done through the intermediary of song: by hearing a lot of singing and a lot of talking they begin to be able to identify which is which. After this process starts they then dissect these sounds further, hearing the difference between lullabies and energetic songs and the difference between regular speech and angry or excited speech.

Next they start to hear the sounds inside speech. This process has fascinated researchers because learning a language is cognitively tough, yet almost all babies and toddlers manage to do it. Also, languages sound and work quite differently and researchers have wanted to understand how the human brain has the capacity to learn languages that are tonal, such as Mandarin or Cantonese, and languages that are based on emphasis or stresses, such as English. It turns out that the auditory processing network is very flexible and picks up the unique aspects of each language, regardless of whether they are tonal or stress-based, to make a huge iTunes library of language sounds from a very early age.

Vowel sounds are the most common in all languages so these are identified first. This may be why you hear babies respond to your parentese with an "ah" or "ee" sound. Then comes consonants, the sounds of which vary more. You can hear this speech development with "da da da" sounds for "daddy" often emerging first (don't take it personally, mothers—just try saying "ma" slowly and you'll see how much more control you need to make the "m" sound).

What happens next is amazing. Babies begin to hear the transitions between vowels and consonants, how our voice and mouth moves from one sound to another inside a word. It is at this point that babies and toddlers make what is called a phonological representation of the sound in their brains, like a brain recording in their internal iTunes library of speech sounds. It has got to be the biggest iTunes library ever and it is all inside those little heads.

At the same time, babies' and toddlers' brains are filling the same library with the *prosody* of speech. Think of this as all the other parts of speech that aren't the words, so the rhythm and melody of speech. This inflection, tempo, and articulation of speech tells us, for instance, if someone is genuinely happy or hiding something from us. These brilliant little brains are also finding the *envelope* of speech, the start and the end of words and sentences. Listen closely next time someone is speaking around you and you'll hear that those gaps are really tiny.

But there is more! Babies and toddlers are also trying out new combinations of words and starting to understand the *syntax* of language, which are its rules and possibilities. All of this heavy-duty cognitive work is being done simultaneously, constantly, and iteratively. There are truly fireworks going off in those cute brains.

Music is so much more than language

While music and language share many characteristics that over time have coevolved, music is the precursor to language in our brains. While acquiring language and being able to communicate is a vital and all-consuming step for babies and toddlers, research has shown us that music might have an even deeper role to play in being human. Just before I start my next story, a quick note about the words baby, infant, and toddler. In the research I am about to share they use the word infant, which is usually a child under the age of one. After the age of one we tend to use the word toddler. In the study I am about to talk about they use the word infant; however, the participants were fourteen months old, which in common language may be referred to as a toddler. This could come from the fact that much of this work comes from infant labs who can work with children from birth to four years old, or because the research works with a term called infant-directed song, when we sing to our babies and toddlers. All you need to be aware of is the multiple uses for the term infant.

In a landmark Canadian study, a team led by Dr. Laura Cirelli researched whether an adult singing to or moving in time with a beat with an infant influenced that infant's helping behaviors. These seem like totally disconnected ideas, but Dr. Cirelli found that singing and moving to a beat with an infant may have contributed to forming a trusting bond between an infant and an unfamiliar adult. Incredibly, this bond later led to the infant both identifying that the adult needed help with a situation and in many cases taking action to help. Her research found that singing with an infant and moving in time with a beat with an infant may have a very primal connection to belonging and contributing to our tribe or family. I was lucky enough to watch this experiment being done with Dr. Cirelli and it blew my mind.

In a nutshell, Dr. Cirelli would interact with the infant (with mom or dad present) in one of the following ways: singing a familiar song, reciting the lyrics of a familiar song, or remaining silent while the child's parent read her a book. Next, she would encourage the infant to watch her complete an activity such as hanging up clothes or drawing with pencils. Then Dr. Cirelli would drop a peg or a pencil and reach for it, saying "oh no" or "uh-oh."

Here was the moment. What would the infant do? Would she notice what had happened or register that it was a problem? Did she understand she might be able to help this unfamiliar adult? Did she take action and actually help this person she didn't know? And did the interactive behavior, such as singing beforehand, matter to the outcome?

I watched this experiment holding my breath. Here in front of me was a test of the capacity to feel empathy and then agency to help another person—and the person being tested was a fourteen-month-old! I thought about myself and how many times I had seen someone trip or drop something, recognizing a problem and maybe feeling empathy but not taking action to help. How much do you admire people who jump in to assist?

The outcome of this experiment, one that has been repeated in numerous studies, was that the infants who were part of the song were far more likely to both see the problem and then take action to assist compared to infants who had not experienced these two preparatory activities. Cirelli has also done extensive study into synchronous movement, moving to the beat of music, with babies and helping behaviors, with similar results.

This research is starting to uncover how music can and realistically has been used for millennia to encourage connection between infants and their family or tribe. This bond through music may be the most powerful way that we as humans develop connection, to the point where it turns into "prosocial" behavior, the ability to feel empathy and then agency to help others in our tribe. It starts early and simply but it may have a more powerful impact on the humans we grow into than we'll ever know.

Rock it out

Music and song are one of our little ones' first language, but I'm not sure we know that. The act of caring for every need of a young baby is overwhelming and all-consuming. Their basic physical needs are paramount and then there is the constantly shifting sands of their development. I always marvel at how a welcome development like being able to walk suddenly turns into a scramble to move everything higher off the ground and to attach bookcases to walls as we wonder why we worked so hard to help them learn this new skill!

With all the things on the "must do" list for healthy baby development, singing to them can seem just another time-consuming and optional activity. But research tells us that singing to our babies is not a luxury but a fundamental need, and to me this reinforces what we have known for generations but might have lost sight of in recent times.

Just sing. Sing *to* your babies, not at them or near them. Look them in the face and maintain their attention for as long as you can. Take control of the other sounds in their environment and take one minute out of each day to sing to them.

Sing to your babies in whatever way and environment works for you. My favorite time with my daughter was the 3 a.m. feed. The house was quiet, "no one" could hear me (although I don't think I would have cared if my husband or dog heard me sing) and I was her favorite rock star. The trick was not to sing too loud as those ears were sensitive information-gathering machines and it didn't take much to turn them on, and I still wanted her to go back to sleep. The next thing was to sing whatever made me feel good, because when I felt good, she felt good. There is a wonderful scene in the now old movie *Three Men and a Baby* where Tom Selleck's character sings a sports commentary to the baby they are caring for. It doesn't matter what you sing, it just matters how you sing it.

Sing to create connection, sing to reinforce safety, and sing to make yourself feel good. Singing has been found to be an effective tool in treating postpartum depression, improving mood, and moderating physiological symptoms of stress. Who doesn't need those things when caring for a baby?

Sing to create empathy and agency. Empathy is a concept that is incredibly hard to teach and even harder to feel, and yet in the world our little ones will inherit, empathy may be more important than ever. Even more important may be the ability to take action to help others and, in a slight update of Edmund Burke's words, the only thing necessary for the triumph of evil is for good people to do nothing. Imagine if we could change the world one song at a time.

Further reading

Adachi, M. & Trehub, S.E. (2012) "Musical lives of infants," in G. McPherson & G. Welch (eds), *The Oxford Handbook of Music Education*, Volume 1, New York, NY: Oxford University Press, pp. 229–47.

Cirelli, L.K. & Trehub, S.E. (2018) "Infants help singers of familiar songs," *Music & Science*, 1, doi 2059204318761622.

Field, T. (2010) "Postpartum depression effects on early interactions, parenting, and safety practices: A review," *Infant Behavior and Development*, 33(1), pp. 1–6.

Harvey, A.R. (2018) "Music and the meeting of human minds," *Frontiers in Psychology*, 9, p. 762.

Nakata, T. & Trehub, S.E. (2004) "Infants' responsiveness to maternal speech and singing," *Infant Behavior and Development*, 27(4), pp. 455–64.

Perani, D. (2012) "Functional and structural connectivity for language and music processing at birth," *Rendiconti Lincei*, 23(3), pp. 305–14.

Sugden, N.A.,& Marquis, A.R. (2017) "Meta-analytic review of the development of face discrimination in infancy: Face race, face gender, infant age, and methodology moderate face discrimination," *Psychological Bulletin*, 143(11), p. 1201.

Trainor, L.J. (2006) "Innateness, learning, and the difficulty of determining whether music is an evolutionary adaptation: A commentary on Justus & Hutsler (2005) and McDermott & Hauser (2005)," *Music Perception: An Interdisciplinary Journal*, 24(1), pp. 105–10.

Trehub, S.E., Unyk, A.M., Kamenetsky, S.B., Hill, D.S., Trainor, L.J., Henderson, J.L. & Saraza, M. (1997) "Mothers' and fathers' singing to infants," *Developmental Psychology*, 33(3), p. 500.

3

THE ENDLESS SONG

Why your toddler just won't stop singing

Me: "Where are we going next?"
Little Miss 3: "We are going up the latergatter to get to the hostipal."
Me: "Yes, we are."

I completely understood what my daughter meant to say: "We are going up the escalator to get to the hospital." But forever more in our family we will call an escalator a "latergatter" and a hospital a "hostipal." Most families have these "cuteisms" from childhood. Sometimes they are the reorganizations of consonants like "hostipal" or a melding of one word with another, such as alligator and escalator. Sometimes there are cuteisms that stick for life. One of my dearest childhood friends has the beautiful name of Siobhan but we always called her Bonnie. The family fable goes that this was because her younger brother couldn't pronounce Siobhan. He could get the *von* from the end of her name, but pronounced it *bon*, and from then on she was Bonnie.

The cuteism period usually occurs when a toddler is really starting to get a handle on stringing words together to communicate. He has mastered the steps of separating sounds that we talked about in the last chapter and is now playing around with the words and combinations to make meaning. But he is still heavily reliant on the musical elements or characteristics, the rhythm and melody of speech, to create these combinations.

In this chapter we will look at how toddlers use song as a transition tool to acquire confident language skills and look inside their brains, as they seem to both incessantly and delightfully sing their way through their preschool years.

There are latergatters on my escalators

Say it quietly to yourself—*es-ca-la-tor* and then *la-ter-gat-ter*. There are the same number of syllables, similar length of sounds, and similar emphasis within the word. To translate this into musical elements, it is the same number of notes with a similar rhythm and corresponding accents (think of accents as the appropriate stresses within the word). My daughter was really close to the correct word, close enough as far as her ear was concerned to convey her meaning. Combine this with her finger pointing to an escalator and that's enough for another person to understand what she means.

Ordinarily you would hear a parent or teacher pronounce the word correctly after Miss 3's attempt and encourage her to try it out herself, to repeat the word correctly. I might have said, "es, es, escalator" to her, to reinforce the change that needed to occur. But in this case, long after she could pronounce it correctly, we keep saying latergatter, just for a laugh.

While the cuteism period is a lot of fun for everyone involved on the outside, inside my daughter's brain there was some serious auditory processing going on. In technical terms, she was making a phonological representation or, put another way, a brain recording of the specific language sound that starts the word *escalator*. She had to hear both the sound *es* and its use in a word, *escalator*, possibly thousands of times in order to make a reliable brain recording of that sound.

What do I mean by a reliable brain recording? Well, think of it like hitting a tennis ball with the same part of a racquet, aiming in the same direction, applying the same force and creating the same intention each time. Think of how many times Serena Williams or Roger Federer needed to practice a serve in order to know she or he could do it consistently, and exceptionally well, every single time. Auditory processing of sounds is a similar process—practice, refine, repeat, thousands of times. There are seventeen different types of tennis shots according to Wikipedia, which seems like nothing when we think of how many language sounds and combinations of sounds our brain is making recordings of when we first learn language.

Nature and nurture

To make these brain recordings of speech sounds, toddlers need to hear them a lot and use them frequently. How frequently is different for every toddler and depends on a myriad of factors.

The first factor is each child's predisposition for auditory processing. A predisposition is in essence a level of ability, capacity, or potential you are born with. In this case we are talking about a predisposition for processing sound, which may have a genetic or familial component to it. A high predisposition for auditory processing is connected with children who we may have previously labeled as musically talented. But in fact, it may also be that they were just using their high predisposition for auditory processing to build their musical talents upon. They might not be musically talented as such, it may just be easier from the beginning because of their predisposition. It is important to remember that there are children who are in a different category when it comes to predispositions for musical ability. Musical prodigies have far above the normal range of predispositions for music and exhibit extraordinary musical skills.

The whole concept of predispositions has challenged many of the ideas we have about human development. Research has previously focused on whether an ability is nature or nurture, that is, if we are born with it or if we can learn it. The concept of predispositions has shifted the research and thinking to look at the contributing factors to any specific ability that are both nature *and* nurture. I like to say I was not born with a likely predisposition to be a great basketball player, mainly because I am five-foot-five. But here is the rub: I might not be tall but I might have had a hidden predisposition to fantastic hand-eye coordination that was never realized because I didn't get the opportunity in life to develop it. The point here is that we are all born with predispositions for many skills and abilities, but those predispositions will only be enhanced or realized if we get the opportunities in life, and particularly in child-hood, to develop them. Both nature and nurture have a role to play.

Which takes us to the second factor that influences how toddlers make brain recordings: through their sound and speech environments. For toddlers to start making brain recordings they need a rich and varied sound environment. They need to hear a lot of speech, and this doesn't just mean people speaking around them: it means familiar and unfamiliar people speaking to them in a variety of different ways and in different environments.

Let's think about this from a toddler's point of view. Her brain is working overtime to try to separate out speech from all the other noises in her world. If you are reading her a book it is ideal to not have any distracting sounds in the background, like a TV or music playing. That is just making her brain work harder. But in today's noisy world, speech very rarely happens without any other noise going on so the toddler needs oppor-tunities to separate speech out from the sound of a truck or a fan or a busy street. The important thing to remember is that if she doesn't catch an instruction or a question straightaway, it could be because, literally, her brain didn't hear you.

Not only do the sound environments need to vary, the speech sounds need to vary. If you dare to venture onto the Internet for parenting advice (it's a jungle in there) you might see tips about surrounding your little one with only positive language and vocal sounds so he feels safe and bonds with you. I absolutely agree that a baby needs to know and trust his carers and that this begins with him connecting with your voice. But as he grows and his brain acts as a language sponge, he needs all the colors of the rainbow when it comes to speech. If music is sound and sound is information, then speech is the gold mine of sound information. Toddlers need everything from joyous, angry, uncertain, and calm speech sounds to feed their brains. Each one of these speech sounds has a different rhythm and melody and this feeds and grows their auditory processing networks.

Less is not more

The reason we know so much about speech development and its connection to music is through studying children who have grown up in disadvantaged or challenging circumstances. Put simply, this is because their language development has been found to be notably delayed or undeveloped, which then leads to problems with literacy skills in reading and writing. Interestingly, when music programs have been implemented with these children their literacy and language skills have improved. But why?

The answer is that music learning trains the auditory processing network, which is also primarily responsible for language learning, and it is the auditory processing network that is underdeveloped in children from challenging backgrounds. A child growing up in such circumstances probably hears less speech and may well be surrounded by more noise in their first five years of life. Their brains haven't had the varied sound environment and the necessary number of speech inputs to create enough reliable brain recordings.

I am a fan of simple sayings that help me remember a concept. In this case it's "If you can't hear it, you can't say it." (In Part Two this will be further developed to "If you can't say it, you can't read it.") A frightening statistic that often features in the research is that children growing up in disadvantaged environments hear 30 million fewer words than other children by the time they are five.[1] That means 30 million fewer speech inputs, and 30 million fewer opportunities to create brain recordings of speech. Furthermore, the speech inputs they hear could be from recorded sources, such as television or radio. These speech sounds may be missing fundamental elements that our auditory processing network uses to make meaning out of speech sounds, and therefore contain less information, or nutrients, for the brain to learn from and grow. I will talk more about this issue when it comes to school readiness in Chapter 6.

So why have music programs been found to have such a profound impact on children who come from these backgrounds and are struggling with their language and literacy skills? Several studies have shown that music learning has improved previously underdeveloped auditory processing skills, and this may have transferred to literacy and reading skills over time. Music learning may have acted as a tool to assist the brain to improve its auditory processing and to process speech sounds with greater accuracy and this is integral to reading success, which we will talk about in Chapter 8.

From music to song to speech

A key factor in many of the longitudinal studies about music learning and brain development is the role of singing. Anyone who has listened to toddlers speak will recognize that singsong style of speech they use, sometimes for months, sometimes years.

---

1   Hart, B. & Risley, T.R. (2003) "The early catastrophe: The 30 million word gap by age 3," *American Educator,* 27(1), pp. 4–9.

Some adults maintain that singsong style and are recognized for it. Often these people make great audio book readers or narrators, like Stephen Fry or David Attenborough. But with toddlers this singsong style is an indication that they are using their music processing networks to practice, listen, and repeat the many speech sounds they are making brain recordings of.

Before the singsong style often comes the maddening repetition of sounds, the "bah bah bah" or "eeee eeee eeee" sounds that seem to be in vogue for three days before a toddler switches to a new sound for the next three days. Little Miss 1½ is bedding down those brain recordings and then moving on to the next speech sound. When she reaches toddlerhood and has just enough speech sounds to get started on phrases, the singsong comes through. As adults we hear it as song, but that is because we have already gone through the process of separating song from speech. Little Miss 1½ is only just starting on that journey.

It is at this point that I often hear parents and grandparents say, "My child is just so musical, she won't stop singing." I hate to burst their bubble, and I must admit that often I don't. The truth may be that their child isn't actually musical in the original sense that they mean it but may be using his music processing network to figure out the incredibly complex cognitive task of learning how to master language. Another theory is that because music and language have evolved from a common precursor (sometimes called musilanguage) they have aspects in common. They could complement each other as he grows by using song to learn speech and he has to first hear it, then make it, then check it with his brain recording to complete that process. When I do stop those parents and grandparents and explain the process I always end with the statement: "So he is musical, but then again, every child is, and you were (and are), too, because you have learned how to speak using language."

Rock stars and role models

Language, when children are young, could be thought of as a combination of rhythm and melody. But imagine if you could put language under a microscope and zoom in a thousand times. What you would see, and what a toddler's brain begins to hear and process, are the micro-sounds of that rhythm and melody. These are the tiny nuances of every language sound, and how they relate to one another is influenced or changed by the other language sounds around them. It is easy to hear those nuances in song because the rhythm is slower than speech and the distance between sounds is larger. That is why song is such an effective tool to teach and learn language.

There is a lot of excellent research and advice on reading to our children so I don't want to repeat that here, but you can look in the Further Reading section of this chapter for some ideas. The key message is to read to them often and in a sound environment that doesn't make their brains work too hard to separate the sounds. There are, however, two practices that I feel we may be neglecting in our communication with children, and they are modeling and correcting speech sounds.

Toddlers learn from us. Sometimes it is far more than modeling—they can expertly mimic us to the point where we might wonder "Do I really sound like that?" We are their speech models in everything: speech sounds, sentence structure, and especially tone. I don't mean the odd swear word or specific accent, I mean in the way we use language. As parents and carers we are their first and most influential language teachers and, as such, what comes out of our mouths goes straight into their ears and gets recorded. When I realized this I became very aware of my language patterns, and if I wanted my daughter and my students to enunciate language sounds well I realized I had to model that for their ears.

The second area that we may be neglecting concerns language sound correction. While we have kept the "later-gatter" alive and well in our family language, we have also corrected our daughter's speech patterns so she can say the word correctly. We know that this will then assist her to read and spell words correctly. It is interesting to look at the current focus on poor levels of literacy learning and wonder if we are missing a fundamental step in the process. If a child can't hear it, she can't say it, and then how can she read it and use it in more advanced written and spoken language? Should what we have learned through studying music be informing the larger issues of learning to use language in all of its forms?

For now, the main thing to remember is that your voice is a song inside your toddler's ears. It is a song filled with information and your toddler is trying to separate out the elements, digest, and process them. Let them sing the same song for days and parrot language sounds until their brain recordings are made. And recognize that you are their rock star and their role model for language.

Further reading

Hansen, D., Bernstorf, E. & Stuber, G.M. (2014) *The Music and Literacy Connection*, Lanham, Maryland: Rowman & Littlefield.

Henriksson–Macaulay, L. (2014) *The Music Miracle: The scientific secret to unlocking your child's full potential*, London: Earnest House Publishing.

4

BEAT BABY

How your toddler connects beat to body

Oh, the joy of taking a toddler to the supermarket. It could end with a happy, laughing little one who is enchanting to everyone who sees him, or it could end up in a facedown tantrum that could wake the dead. As a parent, you never know what is going to happen.

This chapter isn't going to be about tantrums, and I apologize if that disappoints some of you. Instead, it is going to be about the first scenario, the enchanting little one who beguiles everyone with his dance moves—or what we interpret as dance but is actually cognitive growth hard at work.

Supermarkets often have delightfully inoffensive music playing in the background. Songs with a steady beat, often ones you would find on a greatest hits list from the 1960s and that we strangely seem to know most of the words to. I know I am not the only one who has started singing along to "R-E-S-P-E-C-T" by Aretha Franklin just as I have seen those telltale signs of my own daughter getting ready to let out a massive wail.

When a little girl or boy has reached walking stage, sitting in the child seat of a cart often just won't cut it. They want to explore and test out this new walking skill so they are off, often without a speedometer attached. This is when completing the weekly supermarket shop with everything you need and a toddler in tow becomes an Olympic sport.

If you are lucky, somewhere in this marathon your child will become entranced by that Aretha Franklin song and start to "dance." By this I mean that funny little bob toddlers perform when they discover they have knees that bend and try to clap along to the beat, often completely missing their hands coming together. People watch and smile, someone might comment on how cute they are and you have a moment to breathe.

Yes, it is adorable, but there is more going on inside his growing brain than we might have thought. In this chapter we will look at how toddlers use beat to teach their bodies how to move with control and intention, and how music-rich environments help toddlers to conquer the herculean task of learning to control their bodies and synchronize their brains.

What's in a beat?

How do we find the beat in music? Do we feel it somewhere in our body first before that information goes up to our brain, or does our brain find the beat in music and then send it to our body? From the outside one might think the body finds the beat first. We might hear a piece of music, a few moments pass, and then that beat is re-created in our bodies, often with our foot tapping or body swaying. But what happens in those few moments, those few beats, that led to our foot tapping?

The answer is a lot is happening, and it is happening for (almost) all human beings without us even knowing it. Researchers trying to track how the brain might be involved in replicating a musical beat have put forward a number of different models and theories. One by Professor Stefan Koelsch (2011) presents the idea that the human brain, regardless of musical training, goes through an incredibly complex series of interlinked and iterative tasks to decode music before it makes our foot tap. What follows is the short version of his theory, but if you want to go in-depth then look at both Koelsch (2011) and Collins (2013).

First, imagine that as the steps below occur, your brain is simultaneously jumping between the cognitive and emotional meaning of the musical information. Now as that takes place, in Koelsch's model the following processes are happening in consecutive order:

1. The brain extracts all the musical parts. This would be like looking at a whole plate of food, beautifully balanced and portioned, and then looking at each part individually for its features (such as the color, texture, and pattern of the baked pumpkin or the gloss, heat, and coloring of the seared steak).

2. The brain groups these extracted parts together. If we keep with the food metaphor then it would be the brain grouping for proteins or vegetables. The brain also looks

for patterns, combinations, and elements that break those patterns.

3. The brain goes back into its archive of sounds and music and sees what these combinations might be similar to. It is again looking for patterns of what sounds similar or different and in what ways.

4. The brain puts all this information back together and decides if this combination of sounds that makes up the music (the whole meal) is similar enough to be recognizable but different enough to be interesting. This is where the food metaphor might just be stretched a little too far, but stick with me. It is like examining the meal and deciding, by the look of it, whether it is likely to be satisfying and something you will enjoy.

5. The brain sends information throughout the body to embody that enjoyment. It could be tapping, nodding, or swaying, usually all in time to the beat. It also, unbeknown to us, releases brain chemicals that make us feel good and bolster our immune system, so we feel satisfied in the way we would at the end of a good meal.

I once traveled all the way to Norway to meet with Professor Koelsch and he might well raise an eyebrow at the way I have described his model, but he has a fascinating brain that may also enjoy the metaphor. Either way, he has done incredible work in using music as a tool to reveal how our brains learn, grow, and change through experience.

You might be asking how this relates to beat. Well, the beat of the music is one of the aspects that the brain is extracting, examining and then putting back into the music. Other aspects would be the melody, timbre, and rhythm. If you look at the process above, all of those complex cognitive tasks are happening in those few seconds between when we hear the music and when our foot begins to tap. And remember, it is not just those steps the brain is going through, it is also constantly jumping back and forth between the cognitive and emotional meaning that the brain is making from the music. Our brains are also digesting the music for its individual parts as well as the combined sounds at the same time. Put another way, it is like looking at a painting and focusing on the brushstrokes versus stepping back and taking in the whole picture in one go. The amazing thing about the way we process music is that we are doing these two processes, looking at the brushstrokes and the whole picture, simultaneously. Most of the time we have no idea this is all happening, we are just enjoying the music!

Tiny dancer

That cute little dancing toddler's brain is also going through this complex cognitive process, but at this particular age it is possible it is happening with a different purpose. If you watch that toddler carefully, it is unlikely they will be keeping the same beat that you might experience from the music. They will be keeping *a* beat, but not necessarily *the* beat.

The reason for this is that toddlers do not yet have the fine and gross motor control to emulate a beat precisely. But they are trying. Why are they trying? Because at this age toddlers are using music, and more specifically beat, to teach their bodies how to move with control, intention, and synchronicity. This use of music may not be a one-way street—it is possible that the activation of body movement is also helping the auditory processing in that toddler's brain to improve synchronization, which in turn may help synchronization of processes and messages across the brain. I used to see this in my own daughter when one day she couldn't seem to coordinate her limbs to pick up a cup, then somehow the next day she would pick it up like it was the easiest thing to do, something I playfully called a "cognitive click."

One of the most interesting parts of this research is not so much the hitting of the beat but the anticipation that comes before the beat sound is made. Think about our little dancing toddler again. If you watch closely, his movement often comes just after the beat is heard. A beat is a repeating sound with the same distance in terms of time between each repetition, so we can predict when the next beat is going to sound. But toddlers haven't got that prediction wiring all connected just yet and their little dance could be one of the ways their brains are learning how to predict patterns in sound.

Identifying patterns and then predicting the next sound, number, or action in the sequence is the bread and butter of the first years of schooling. As parents we probably see it most in the math activities our children are doing: counting in tens, number skipping in odds and evens, up to subtracting an odd number to get an even number. But patterns are everywhere in our lives and the ability to identify a pattern and then identify the "which one is not like the other" is the basis of all learning, not just math or music.

A great deal of research is currently underway to understand how our brains learn to predict what comes next, what needs to happen next, or how we can change the pattern through a particular choice. Think about our adult lives. How often do we need to first see a pattern (always paying our bills late), identify the problem (they all come in at different times of the month but you only pay them on one particular day, meaning some of them will always be late), and take action to change the pattern (put the mid-month bills on direct debit and pay the end-of-month bills through your old method)? The ability to complete these complex cognitive processes as adults may have started with mastering the ability to predict when to lift up your foot as a toddler in order to bring your foot down on the beat of the music.

This research into beat and how our brains use it to learn and grow has exploded since the mid 2000s. The reason for this is that the concept of rhythm within the brain has opened up a whole new realm of understanding about how the brain functions. The external representation (foot tapping) of an internalized beat that comes to our ears and into our brains as music is an outward indicator of internal brain functions. Put another way, it is a behavior seen outside the body that can tell us something about what might be happening inside the brain. Music therapists and teachers have often observed that a child struggling with their learning will also struggle to keep a steady beat. We now know that this is an external indicator of an internal problem with that child's brain function. But is it a problem that can be remedied? (Now where did I put my drum?)

Old dog, new tricks

Anyone who has watched that little dancing toddler grow will know that, typically, he ends up being able to find the beat to a song. This is because most children are typical learners and being able to tap along to the beat in a song is just a "cognitive click" in their developmental road. Some children learn it through active instruction, which can be anything from formal toddler-focused music classes full of clapping- and walking-to-the-beat games to play with parents or siblings where they sing, clap, and bang pots and pans on the kitchen floor. Some children learn it through having a music-rich environment and finding the beat themselves. Some children don't get any of these opportunities, and this is where research is highlighting how a lack of musical play and beat-keeping games can have a detrimental effect on a child's brain development.

The ability to learn these skills is not just due to exposure to musical activities or a music-rich home environment. It also has a lot to do with the way our brains were wired from birth. Research into brain development over the last twenty years has forced a reexamination of a lot of the ideas we had about the way the brain functions. The concept of predisposition is one of the big ideas that we talked about in Chapter 3. When it comes to beat, it may be that two toddlers dancing in a supermarket may have been born with different levels of predisposition for finding it. There is no beat center in the brain, but there is growing evidence of an intrinsic sense of beat through our neural circuits that have to connect in order to hear the beat, move to the beat, and then predict when the next beat is coming. There is therefore the potential for that process to be achieved in a quicker, easier, or less directed manner by one toddler and not the other due to the more effective neural wirings that she has been born with.

Neuroplasticity, or the ability of the brain to change, is another big idea that has been the subject of much recent research. We have gone from the idea of a fixed brain capacity, as embodied in the phrase "you can't teach an old dog new tricks," to the idea that our brains have the capacity to learn and change throughout our lives. This same idea lives in the concepts of growth mindset versus fixed mindset, which you will see in the title of every third book on the self-help shelf in your local bookshop. A lot of what we know about neuroplasticity has come from the study of musically trained brains. Why? Because musicians were found, even early on, to have higher levels of neuroplasticity and researchers wanted to know why.

When it comes to our tiny dancer successfully bobbing to the beat, it is a combination of the neural wiring they were born with and the experiences they have had in life. As a toddler moves into their preschool years they still may be struggling to keep the beat, but it is important to understand that keeping the beat is not solely a talent or a learned skill—it is both. As we will see in Chapter 8, the ability to keep a beat is a foundation stone for many other learned skills, including reading.

Find that beat

Almost every culture has songs that incorporate clapping in a way that requires toddlers to predict where the clap should be placed. One of my favorites is "Bingo," which is often taught in schools. There is a whole verse about a dog called Bingo but the real heart of the learning comes in the chorus when the children leave out a letter of the dog's name and often clap instead. If I haven't already created an earworm for you by mentioning this song, the chorus would go *clap* I-N-G-O, and then *clap* *clap* N-G-O. To clap at the right place requires an internal sense of the underlying beat in the song, memory of the song structure, preparation of the hands to clap, and movement of the hands just prior to the beat in order to make the clap sound in exactly the right place. And then, as if that accomplishment wasn't enough, you have to keep changing the pattern the next time the chorus comes around. It's tough cognitive work for a three-year-old, but if you ever get the chance, stand back and watch a group of children trying to accomplish this. It is a fantastic diagnostic tool for looking into their developing brains to see which ones have the neural connectivity in place to achieve this seemingly simple feat.

Clapping is simple, enjoyable, vital, and seems like it takes forever to master. Your little one needs to watch you clap along to music, they need you to hold their hands and help them to clap in time and it needs to happen every day. When I observe parents in cafes where I write, I am always astounded at how much of their movement with their toddlers contains beat. They pick them up and often mothers will sway in time to an imagined beat without even knowing it. A father might tap his leg along with the music and his little boy, who might just be starting to walk (oh, the trouble that lies ahead), will grip on to his leg and tap, too. These are all opportunities for beat learning and the experience of auditory and motor cortices connectivity, all while you get to enjoy a coffee and maybe some cake. The most important point in all of this is to notice when beat-keeping is being learned and how we as parents can enhance that process.

At times we might inadvertently shut down that beat-keeping process. Sticking with the cafe scene, sometimes I have seen a toddler kicking a table leg in time. It is great fun to them because it is both intensely tactile, their foot is hitting something hard that might also wobble, and it is often auditorily satisfying because it makes a big noise. Culturally it is not really acceptable because it is disturbing other patrons at the cafe, but cognitively it is fantastic.

We will often stop Little Miss 4 from making the noise and even admonish her for her actions when what we really need to find is a balance between a cognitively rich experience and cultural expectations. I am not saying we should let Little Miss 4 continue making noise and announce to the other patrons that she is connecting her auditory and motor cortices with a complex cognitive process, so just deal with it. I am saying with this new knowledge we have the ability to recognize an important cognitive click that we need to encourage but also channel into a slightly quieter avenue. Challenge her to keep the same beat on her knees and then cross her hands over every second beat. It will keep her going for hours.

Further reading

Collins, A. (2013) "Neuroscience meets music education: Exploring the implications of neural processing models on music education practice," *International Journal of Music Education*, 31(2), pp. 217–31.

Koelsch, S. (2011) "Toward a neural basis of music perception: A review and updated model," *Frontiers in Psychology*, 2, p. 110.

Phillips-Silver, J. & Trainor, L.J. (2007) "Hearing what the body feels: Auditory encoding of rhythmic movement," *Cognition*, 105(3), pp. 533–46.

Trainor, L.J. (2007) "Do preferred beat rate and entrainment to the beat have a common origin in movement?," *Empirical Musicology Review*, 2(1), pp. 17–20.

5

I SEE, I SAY

Getting your child ready to read with music

A few months into my PhD, a colleague for whom I had a great deal of respect said to me, "About halfway through your PhD is when you'll figure out the real question you are trying to answer."

I dismissed her comment almost straightaway. I had just spent months drafting and redrafting my research questions with my supervisor, choosing every word of every question and sub-question so carefully. I knew the questions that my PhD would answer; they were all there in black and white.

How wrong I was. Actually, I wasn't wrong, I just misinterpreted what my colleague had said. When she spoke about the real question I was trying to answer, she didn't mean my research question—she meant my life question. As you will see in the coming chapters about learning to read, I think my PhD and all the work I have done since in the field of neuromusical research has been trying to answer one simple question: Did learning how to read music help me to learn how to read words? This chapter will follow my own personal story to show how music and language reading can strengthen our brain connectivity in this tricky cognitive process.

The great pretender

My simple life question needs some context to make sense. I often use my own story to explain how music learning has the potential to change the course of a child's life. I am the firstborn child of an elementary school teacher. Later in her career my mom trained to become a specialist reading teacher, helping children who were struggling with their reading to find ways to decode written language. She was exceptionally good at this and changed the lives of many children. I think she was so good at it because she understood that the task didn't just involve identifying and remedying each student's specific learning issue but was also about rebuilding the student's confidence and self-image as a learner. A child's confidence gets bruised and undermined in our Western educational system if he or she can't read by eight years of age. There is an educational adage that encapsulates this: from the ages of five to seven children are learning how to read, but after the age of eight they are reading so that they can learn.

Picture what it is like at home every night during reading time when the firstborn, chatty, and headstrong daughter of a reading specialist teacher is struggling to read. Here is a teacher with all the tools you could ever need to teach a child to read, and she can't figure out why her own daughter can't read fluently. As a mother myself I actually can't imagine the stress this would cause.

I will add that I didn't help. I was channeling my inner teenager at the age of seven and didn't want help from my mom. My attitude was "I can read, don't tell me I can't." Now when I look back I can see that I could fake-read very well. What I mean by that is I had a good memory and an even better ability to fill in the gaps in a sentence. I would study the shape of the sentence, look very hard at the picture beside the text, pick up nonverbal clues from the person I was reading aloud to and then manage to get close enough to what was actually written that the grown-up would just correct one small thing, thinking I could read. I wonder if my current role as a musical conductor came so easily because I was an early master at reading the nonverbal signal.

I also tried to surround myself with very smart students. In the later part of elementary school I had a close-knit group of friends (I still have them today) who were topping tests and winning academic prizes. When actors say they surround themselves with other creatives at the top of their game so that it lifts them to new heights, I know what they mean.

In about fourth grade, when I was nine years old, my tricks were wearing thin. The content we were reading was getting longer and more complex and I couldn't keep up. I couldn't hide anymore and I started to develop a number of unattractive defensive behaviors in a vain attempt to manage the issue. My internal confidence was also suffering very badly and to this day I struggle with the feeling that I'm not measuring up or smart enough in whatever aspect of my life it may be.

At the very same time that I was spiraling downward, a music program in the form of learning an instrument in a concert band was offered at our school. My very smart friends and I were all selected to be involved. I still remember holding a piece of paper that had my name on it, a whole bunch of writing I couldn't really read, and the word flute handwritten (which I really only knew because the teacher said the word as she wrote it down). I asked for a flute when it was my turn at the front of the line. The teacher informed me they had just run out, but how would I feel about a clarinet? I was so excited to just be selected I would have taken bagpipes if they were offered. I walked home that day with a clarinet, my clarinet, in my hands.

What happened next is the basis of my simple life question. I was taught how to play my clarinet, which it turns out was a far better match for me physically than a flute (I have short fingers with wide finger pads and a whole bunch of hot air inside of me). I loved it. Here was something, finally, I could do with a measure of success that was at or above the level of my peers. And it was something that I got better at every time I practiced it, not like reading where I just seemed to get worse.

From the very first lesson I was not just taught how to make a sound (or a squeak) on my clarinet, I was also taught how to read music. My teacher showed me the score as I held the instrument in my hands: this is a "D" and to play a D you put down your thumb and first and second finger on your left hand and blow into the mouthpiece. With that simple experience I now believe that something happened in my brain. I was connecting a symbol with a sound. Musical notation wasn't as complex as words and it also didn't have confusing rules that you both follow and then break as you have to learn to do when reading English. The note D was always a D in the score.

It didn't happen overnight but about six months after I learned to read music I was finally really able to read words. I started to read more fluently, I didn't break into a sweat and feel like I had an instant fever every time I was asked to read out loud in class and I actually enjoyed reading before I went to sleep, rather than scanning the words and turning the pages so it looked like I was reading. Don't get me wrong, I still had and have problems reading, but something was permanently, and fantastically, opened up in my brain.

A path interrupted

Looking back at this experience now after twelve years of schooling and a further twelve years of university study, I think learning an instrument and how to read music dramatically changed my life path. What would I be doing now if I hadn't had that opportunity to rewire my brain and reverse the decline in my confidence as a learner? I honestly believe that the reason I have done so much formal study and taken on a ridiculously difficult area of research is to continually prove to myself that I can read.

Back to the start of my doctoral studies, it took three more years, the completion of my PhD and a trip to Evanston, Illinois, for me to find the answer to my real-life question: Had learning to read music helped me learn to read words?

The symbol to sound system

After my PhD was completed but before it was awarded and I got the "Dr." in front of my name, I applied for a travel scholarship to visit the research labs I had learned so much about during my studies. One of them was the lab of Professor Nina Kraus at Northwestern University in Evanston. It is a strange experience to send an e-mail to one of your research deities who doesn't know you from a bar of soap and ask if you can come and camp in their lab for a week and just learn. I also find it astonishing that every one of the research labs I have asked to visit, and I am at close to twenty now around the world, has said yes. I can never thank them enough for their openness and generosity.

Brainvolts, as Professor Kraus's lab is known, is an auditory neuroscience lab that researches the auditory processing system. They use what they call a biological approach to understand brain development and also study learning delay or impairment. Professor Kraus calls our auditory network the window into our brain.

While I had heard of the phonological loop, or the symbol-to-sound system, through the work of Professor Aniruddh Patel at Tufts University, this was my first opportunity to speak with auditory neuroscience researchers about my simple question. While there I also undertook many of the scientific tests that the participants perform. I wasn't sure what to make of one of the researcher's comments when looking at my results: "Well, you have an interesting brain."

Through a number of interviews with Professor Kraus and her team, I started to distill what is a very complex cognitive process, also known as the phonological loop, into simpler steps. From here I could then overlay my own experience of learning to read music and words.

The phonological loop

Below I set out the steps for the phonological loop. I am presenting it in its simplest form and I have deliberately chosen to not include some of the more complex sub-steps in this list.

1. First the eye sees a symbol on a page. That could be an eighth note "D" in music or a "t" at the start of a word.
2. Next the brain hears that sound. The best way to describe this again is as a brain recording held in the very large iTunes library of music and speech sounds we each possess.
3. The brain tells the body how to make that sound. On my clarinet it is as I described earlier (thumb and first and second finger on left hand down and blow into the mouthpiece). For a "t" at the start of a word my brain tells my tongue to go to the top of my mouth, put air behind the sound and then release my tongue.
4. Then we make the sound, either through the instrument or our speech.
5. Immediately we listen to that sound and check for matches with the sounds in our personal library. Does it sound like the brain recording or does it sound different? Depending on the answer, do I need to reinforce the brain recording or change it?
6. And repeat.

Think about how quickly this process unfolds when you read a news article or an e-mail, or when you watch someone perform a piece of music as they read it from a musical score. It's superfast!

My interesting brain

It would be nice to be able to point to one of these steps in the phonological loop and say this was where my own brain wiring was faulty, this was why I had difficulty learning to read words, so let's just fix that. But it isn't that simple, and it won't be for many children who struggle to read. It could be that my first reading of the symbol was corrupted, or my brain recording was not reliable. Maybe I couldn't make the sound correctly, but by all accounts I was a very articulate child. There is also the important emotional reaction I had when I tried to read that is likely to have impacted my brain wiring. The same researcher who told me my brain was interesting said that brain imaging and testing today was unlikely to give me a definitive answer about why I have difficulty reading, because I have subsequently rewired my brain to compensate for my initial problems. The answer is that I will never know if the connection is causal—if my learning to read music allowed me to then learn to read language—but as I mentioned I do know that learning to read music and play an instrument changed my life path significantly.

Not every child learns music, but they still learn how to read

A large number of children never learn to read music but the majority learn how to read words. This chapter isn't meant to challenge or change the way we teach reading; that is an area for the language and literacy experts around the world.

What I do want to highlight is our new understanding of how the study of music reading has deepened our understanding of language reading and how learning music could be viewed as a complementary learning tool to language learning.

From a music learning perspective, it is vital that the element of symbol-to-sound connection is included from the beginning. I believe this is vital for two reasons: to ensure that the capacity for music reading to support language reading is utilized, and to allow children to develop effectively as musicians. It is not uncommon for children to learn music without learning to read music, and although it is a pillar of the Suzuki method, I believe the research points to the benefits of music reading being initiated as soon as appropriate for every child. The symbol doesn't need to be complex—indeed, many music education methodologies for early childhood use a line and a squiggle to indicate the sound of silence. Next time you see your child in a music education class, look for the symbols they use—that is your child's brain doing a pre-literacy workout!

We shouldn't leave this process to the once a week music lesson, although this formal learning experience is important. The more important first step is to notice when our children are connecting what they see with what they hear.

My favorite example is an emergency vehicle. Children are fascinated by fire engines, ambulances, and police cars, and why wouldn't they be—they are big, they have flashing lights, sirens, and they go really fast. Children see the emergency vehicle first and hear it second. Then they might point and say, in their pre-language babble, "ire-ruck" for fire truck. At some point this can switch around and the child hears a fire truck, knows what it is, and then tries to see it. If we think about this in cognitive terms, the first example is the visual processing in the brain talking to the auditory processing and making a connection. Children need quite a few experiences of that connection to make the connection strong enough to go the other way, from their auditory system through their cognitive/memory networks to their visual network.

I use this example to point out how many repetitions children need of simple connections to make it mean something and join the cortices together. As adults we are highly evolved learners and often don't need to see things too many times to then commit them to memory and create a new neural connection. We are building on a whole city of neural connections, so adding in a new road isn't hard. It's not the same for children: their brains are like an open field with a few hillocks and maybe trees but no structures. Their experience builds the roads and buildings and we as parents are the high-powered steering committee for the construction.

Repetition is vital when making the visual/auditory connection between symbols and sounds. I always love the "follow the bouncing ball" idea used to teach songs on *Sesame Street*. The words are displayed across the bottom of the screen and the bouncing red ball moves in time along them as the children or characters sing along. This is connecting the symbol of the words with the sound of the words with the rhythm or movement of the words over time. The interesting thing about reading for every child is they need many and varied experiences of connecting symbols to sounds.

When we read picture books we often sit so that our child can see the pictures while we read the words. How different would this experience be if your finger followed along the words as you said them, and you spoke a little slower so the child could make that connection? I did this one day with a little girl at my daughter's day care. In true mimicking style, she pushed my finger off the words and the book and started following them herself. When I stopped talking in the middle of a sentence, her finger stopped. She was connecting the symbols to the rhythm of my sounds.

When my daughter came home from school with her first PM reader—a widely used reading program of leveled books (1 to 30) that aims to sequentially build children's reading skills—there was also an instruction sheet. Now, you can probably imagine that, considering my own history with reading, I was very anxious that my daughter's journey to fluent reading be far smoother than mine. On the set of instructions for the reader, one dot point said, "Wait for your child to figure out the word in their head and try not to interrupt with the letter sounds." Couple this instruction with my anxiety about my daughter's reading journey and my own tendency to go too quickly and this felt like a herculean trial. But I did it, and as I sat with her I wondered if this was the symbol-to-sound system in slow motion. She was seeing the symbol and then searching her small but developing iTunes library for her brain recording of the sound. Then her motor cortices were talking to her mouth and breath and telling her how to make the sound. The sound came out and, like me, she was looking for nonverbal confirmation that it was correct before giving her brain recording a big tick or a making a new recording of the corrected sound.

This process takes a fraction of a second (sometimes longer, which is when I needed to take a deep breath and find my well of patience) but it is a precious second. Now that we understand more about this complex cognitive process it is like a little miracle taking place with every word they read.

My daughter's reading is fine, by the way; she often brings a book with her to breakfast and she is not just looking at the pictures.

Further reading

Bonacina, S., Krizman, J., White-Schwoch, T. & Kraus, N. (2018) "Clapping in time parallels literacy and calls upon overlapping neural mechanisms in early readers," *Annals of the New York Academy of Sciences*, 1423(1), pp. 338–48.

Corrigall, K.A. & Trainor, L.J. (2011) "Associations between length of music training and reading skills in children," *Music Perception: An Interdisciplinary Journal*, 29(2), pp. 147–55.

Moreno, S., Marques, C., Santos, A., Santos, M., Castro, S.L. & Besson, M. (2008) "Musical training influences linguistic abilities in 8-year-old children: More evidence for brain plasticity," *Cerebral Cortex*, 19(3), pp. 712–23.

Patscheke, H., Degé, F. & Schwarzer, G. (2016) "The effects of training in music and phonological skills on phonological awareness in 4- to 6-year-old children of immigrant families," *Frontiers in Psychology*, 7, p. 1647.

6

SCHOOL READY

How music helps children prepare for big school

Taking on the sensitive topic of school readiness is scarier than giving red cordial to kindergarteners. There is a lot of advice available from educators and parenting experts about preparing your toddler for their first big-school experience. Should they have been to preschool or not, should they be reading yet, and what kinds of social experiences should they have ticked off before walking through the school doors on day one? What does it mean to be school ready? Ready for what, exactly? Advice comes thick and fast and often it is difficult to determine which is useful.

Even as an educator with more than two decades of teaching and research experience, I often struggled to separate helpful from unhelpful advice and to also separate my opinion as an educator and a parent, which did not always match. Ultimately I came to this place: while parents are their child's first teacher, they are not the only ones their child will learn from, and many times will not be the most important. But parents and teachers are essentially doing the same job, just from different perspectives, helping to produce happy, independent, and productive human beings. There are a million different ways to get there and every day and every experience is a new page in that child's book of learning.

With that idea in mind, I became a curious observer of the school readiness process, in terms of individual children and how research into music learning both supported and challenged the way we parent, the way we teach and the way we define school readiness. I don't want to disappoint you so upfront I'll say this chapter won't contain a list of activities you can tick off and then feel confident that you have done everything to help your toddler be school ready. I will, however, look at five areas where the research into music learning can be applied to that magical period just before big school when they are preparing for the exciting and the unknown.

Safe with song

Children use singing to process meaning from the speech of their carers and to determine if a new person is to be trusted. Song helps a child feel part of a family or tribe and it is the language that soothes them to sleep. It is their first language of safety, security, and love.

Not every child gets to feel this connection, through song or other means, from the start of their life. This lack of felt connection and security can fundamentally affect a child's cognitive development and we are beginning to understand the different and often delayed neural development that comes with a lack of physical and emotional safety at the beginning of life.

The reason to go down this quite dark path is to understand how important the first five years of life are to neural development. Music processing and learning can play a part in everything from encouraging typical neural development to bringing about an improvement in neural development if the child has experienced difficult life circumstances in the first years of life.

Being sung to and singing with others has a profound effect on all humans at any stage of life, and it is both easy and totally delightful to see this on the smiling face of a toddler. Making music through singing increases some brain chemicals such as dopamine, changes the levels of some hormones related to dopamine, and reduces cortisol, which is a stress hormone. Amazingly, music does this incredibly quickly and almost acts like a circuit breaker for the production of cortisol in our bodies.

A wonderful story was related to me about an experienced principal and a student who was having a seriously bad day. This principal can handle anything—she has pretty much seen it all and runs a tight and happy school ship in an extremely challenging area. But on this particular day she had a young girl in her office who was intensely angry. The principal was trying everything in her extensive book of behavior management approaches but the anger was only increasing. In a moment of what might have been intuition mixed with desperation, she started to sing. Quietly, she would probably say badly, but still, she was singing. The spiral of anger ceased almost immediately and a magical calm came over them both. This young girl did not need to be told what to do, to be lectured to, to be questioned—she needed to be sung to.

What might have happened in this moment? I don't know for sure but based on the extensive research into how song and singing impacts on our brains and bodies, being sung to made her feel safe in a space where she had lost both internal control and her sense of connection. This experience reminds me of studies that have compared the time it takes for a baby to fall asleep if they are held and sung to, just held and rocked, or placed in their cots and sung to. The findings were that babies fell asleep fastest and slept best when they were placed in their cot and sung to. There is a power in our singing voice.

At big school, song can play a vital role for child, parent, and teacher. Some children self-soothe with song; if you get the opportunity to spend time with a class of kindergarten students you will hear them singing to themselves often. In this phase where language and song are still overlapping as children perfect their auditory processing, teachers will regularly use song to support class cohesion and the felt experience of their new tribe.

From a parent's point of view, that magical time just before big school is a good moment to check where the song is in your child's life. Do they sing alone or with you or their child-care group? When they do sing, do you comment or join in? Is there enough clear auditory space without TVs and other music in the background for them to create their own songs? As a parent you may not have to actively encourage singing but simply make space for it to happen, and then sit back and enjoy it when it does. All we need to do as parents and educators is to notice singing and understand what might be happening inside our child's brain and body.

The attention span of a gnat

The attention span of a four-year-old is something to behold. It can be as fleeting as a fly buzzing past or as extended as a binge of *Peppa Pig* episodes. When it comes to school, attention span is in many cases the first and most important ingredient of learning. You cannot learn if you can't focus, find meaning, and then create memories for more than a few moments at a time.

How many of you, dear readers, had a comment on your school report like "could focus more" or "struggles to maintain attention"? Being able to concentrate on a given task is a skill that develops over time and it is, to a great extent, a learned one. Like many skills, we learn through modeling by people around us, direct instruction ("why can't you just pay attention?"), and our own predisposition. It is a mysterious mixture of influences, but any teacher will tell you that students who can maintain their attention have a much better chance of being successful at school.

How does music learning help? Well, structured and sequential music learning activities at this early childhood level are basically boot camp for the development of any four-year-old's attention skills. A three-minute music learning activity will engage multiple senses. It might involve a child moving his body while listening to a song or beat, watching out for when the song stops, and then performing a new movement like turning around. Or a child might miss a word in a song and then carry on. Music activities require hyper-attention across multiple senses, following varied instructions without any warning while working within a group and couched as a game so the child is having fun. I have done exactly the same activities with preservice teachers and they have commented on how much attention this fast-paced brain workout requires.

Furthermore, the activities can change quickly. The first activity could be clapping to a beat, then three minutes later it might be singing a song with puppets. This rapid change is great for attention development because that flighty toddler has to adjust his attention to a new activity, new expectations, new movement, and a new set of concepts to watch out for. I regularly hear comments from parents that their son can't pay attention for twenty minutes to finish his dinner but he will remain hyperalert during his sixty-minute music class and ask for more at the end.

So often, too often for my liking, parents involve their toddlers in formal music learning classes or choose childcare centers that include daily or weekly music activities in their programs because they want their toddler to be able to express themselves. While I don't deny that musical activities may be able to assist in self-expression, at this important stage in a child's development we now understand that what structured and sequential music learning activities really provide the foundation for is the important school-ready skill of paying attention.

The rain in Spain falls mainly on the train?

I was brought up on musicals. It was my special outing with my parents and something that I got to do just with them, without my younger brother. At the time the most prolific performance troupe in my city was an amateur group, so I have seen every Gilbert and Sullivan musical there is (I have a love–hate relationship with *The Pirates of Penzance*) and multiple versions of *My Fair Lady*. There is a very well-known line in that musical when Eliza Doolittle, a flower girl with a thick cockney accent, is being taught to speak "correctly" by Professor Henry Higgins, an expert in phonetics, by repeating the phrase "The rain in Spain falls mainly on the plain." I had to laugh once when I tried this as an experiment with a toddler who repeated this line back to me and replaced *plain* with *train*. She then asked, "Did they get wet on the train?"

The point I want to make is that the ability to hear the nuances of sounds, the difference between a "p" and a "t," is a vital skill as every toddler gets ready to go to big school. Distinguishing sounds is a foundational building block for learning how to read, one of the most important tasks in the first three years of school. Researchers have now been able to track the phonological process neurologically and the ability to distinguish sounds is the first and most vital step. If your child can't hear speech sounds, and correctly and reliably distinguish between the sounds, then they can't verbally produce the correct sounds in turn. If they can't produce the correct sounds they will also struggle to read.

It is in a child's speech that we can help correct what they hear and effectively prepare them for reading. Working in elementary schools that cater to students from challenging or disadvantaged circumstances, I have been struck by how poor their enunciation is, perhaps twelve to eighteen months behind where I would expect their speech skills to be. This could be due to a number of environmental factors but the most reported one is a lack of direct speech or speech inputs before they enter school. They have just heard fewer words.

Literacy programs in such schools have recognized this delay and focus heavily on oral language development before the students begin learning to read. These programs embrace the hear-speak-read process. Unfortunately, however, the students have already started school and are statistically likely to remain below national benchmarks for literacy throughout their schooling.

Research into music learning has helped us understand just how important auditory processing of speech is to a child's brain development, and yet I wonder how much we focus, either at home or at school, on the ability to enunciate speech effectively. In a world that is noisier than ever before it is important to consider how we manage a toddler's sound environment for their best auditory development.

This does not mean schools and parents have to focus on pronouncing everything in the "Queen's English" or get children to wear noise-canceling headphones and only allow them to hear classical music. Toddlers and young children need sound information in order to develop their auditory processing networks and a lack of noise would be like wanting big muscles without ever lifting up a weight. What we should be aware of is the level of noise in terms of volume, complexity (how many different sounds are being processed at once) and consistency. If we as parents always speak to our young children with the drone of the TV in the background, then that is a higher processing load than muting the TV for a moment and facing a child as we speak to them.

Variety, not the lack of noise, is the key. Exposing toddlers to speech in both quiet and noisy environments and exposing them to speech sounds in other languages, dialects, and accents is a workout for the auditory processing network and hones their speech-in-noise skills. Toddlers who participate in music learning activities have been found to exhibit higher levels of speech sound processing, mainly because, like attention boot camp, music activities are fine-tuners for the sound processing brain. In our noisy world, music learning may be more important for our toddlers' development than ever before.

And let's not forget the other parts of speech that music learning assists. Prosody, the rhythm and melody of speech, is just as important and can be enhanced by music learning activities. I have often heard parents say "Don't talk to me like that, it's rude" to Little Mr. 4 when their previous vocal interactions with their child have used exactly the same tone. This is particularly important in a stress language such as English but may be different in a tonal language such as Mandarin or Cantonese. What music learning can do, as well as maybe offer more effective modeling, is help children to hear how to change the rhythm and melody of music and then transfer that adaptability to speech.

Keeping a beat

I'm about to share with you my favorite piece of research. It is my favorite because it is both profound and immediately applicable in schools and could change national literacy levels in one musical swish. Children need to be able to keep a beat to be ready to learn how to read.

Why, I hear you ask. The answer comes from an interesting observation of children who struggle to read due to dyslexia or some other form of learning delay. Those same children struggle to keep a steady beat, and neuroscientific and audiology processing researchers have wondered in the past if there is a connection. It turns out that there is. Being able to keep a steady beat is an indicator that learning to read is ready to happen in the brain.

A child's brain is required to develop a certain level of sensitivity and synchronization to timing cues in language in order to learn how to speak, which then leads to learning how to read. This is called the pre-literacy or pre-reading phase and involves the brain being able to process speech sounds (phonological processing), remember speech sounds (auditory short-term memory), and name the speech sounds (rapid naming). Interestingly, children who are able to demonstrate beat entrainment (being able to keep playing a steady beat) were found to have this higher level of neural synchronization. Indeed, researchers ended up classing the students as synchronizers or non-synchronizers and found that synchronizers had higher pre-reading skills on all measures.

The reason I like this research so much is that it was conducted with children between the ages of three and four years. While the synchronizers are well on track in their pre-reading skills and school readiness, this age range gives us as parents and educators about a year to help the non-synchronizers get their brain in sync, barring other developmental issues. All they need is a drum, some clapping games, and both an experienced music teacher and a family who is willing to do a whole lot of singing, clapping, and moving to the beat on every car trip they take. Let's throw a Saturday night dance-a-thon in there, too, as long as it focuses on the beat. It is such a simple intervention for such a vital learning step. It won't work for every child—I know there was a lot of singing and clapping in my household and I still have an "interesting" brain—but if it works for most then that is very worthwhile.

The social soup

The other very significant challenge that greets most children when they go to big school is the social soup. Usually there are many more children than they have ever interacted with before in family, day care, or preschool situations. There are extra rules about how children are expected to behave, and at least a few of these won't be the same as the rules they had to abide by at preschool. It is truly a brave new world for them.

Entering into that school-ready phase, the ability to control reactions and interact positively with other children can be a challenge. One of the mechanisms that children are using in this situation is inhibitory control, which helps inhibit our desire to go with a natural or habitual response. In other words, it's the mechanism that stops Little Miss 5 from acting on the internal voice that says, "I want that book, I will take that book." Inhibitory control is hard at work in social as well as learning situations and crosses over heavily with the ability to maintain our attention when we would rather be pulling someone's ponytail.

In many cases, strong inhibitory control is developed through delayed gratification. I think kindergarten teachers must get more than sick of saying, "No, Little Miss 5, Little

Mr. 5 has that book now and you need to wait your turn to look at it." This skill is more than just following the rules and behaving well—it has a profound effect on children's ability to make solid and positive friendships, some of which might last, if you are lucky like me, for life.

Music learning activities are another great tool to teach the "wait your turn" skill. Watch any circle rhythm game and you will see children waiting their turn for what seems to them like an eternity (maybe two minutes at most) but it is flexing their "wait your turn" muscle, which can then be transferred to the "my turn to speak" and "don't strike out when you get angry" muscle. Musically trained children have been found to have particularly high inhibitory control. This can be very helpful in their teenage years when handling emerging challenges such as taking calculated risks, managing stress effectively, and avoiding addictions.

Ready, steady, school

Music learning activities have been part of child development pretty much forever. They help children find their tribe, know who to trust, feel love and connection, develop the ability to understand speech and then communicate themselves, and wire up their brains effectively, positively, and strongly. From the research that has now been undertaken and the new understandings we are beginning to uncover about music learning and brain development in the first five years of life, we can see how helpful it is in so many areas. We won't see many of these benefits until later in their childhood or into adulthood. As one newspaper headline I read said, "[Music Learning] Is the Gift Your Parents Gave You That Just Keeps on Giving." I would add to that "and it is fun, it can't hurt, and it will probably help in ways we haven't even discovered yet."

Further reading

Hansen, D., Bernstorf, E. & Stuber, G.M. (2014) *The Music and Literacy Connection*, Lanham, Maryland: Rowman & Littlefield.

Henriksson-Macaulay, L. (2014) *The Music Miracle: The scientific secret to unlocking your child's full potential*, London: Earnest House Publishing.

Kraus, N. & Anderson, S. (2015) "Beat-keeping ability relates to reading readiness," *The Hearing Journal*, 68(3), pp. 54–56.

PART TWO

THE FIRST FEW YEARS OF SCHOOL

5 TO 7 YEARS

7

THE MUSIC OF LANGUAGE
An overlapping network

"If a child can't hear it, they can't speak it, and if they can't speak it, they can't read it." This was one of the gems of understanding that a researcher gave me when she was explaining how music and language were connected.

This phrase fell out of my mouth one day when I was working with a group of school leaders. I had just presented a whole series of fancy slides painstakingly put together to walk them through the research into music and language development for children but was getting very little response. School leaders, who could be a principal or head of school, to a member of the leadership team who could be interested in anything from leading curriculum to school finances, to pastoral care and well-being, are a tough crowd. They have seen it all, and their minds are only ever about fifty percent with you at any given time as the other fifty percent is trying to solve the multitude of issues playing out in their own schools at that very moment. They will often talk about the tension between knowing professional learning is essential and the difficulty in pulling themselves away from the urgent day-to-day issues of their school. It is no wonder the mental health and well-being of school leaders is at the top of the agenda at most conferences.

They are also a tough group to get on board. Comedians often talking about "dying on stage" when no one laughs and the mood changing in an instant when the first heckler decides to criticize their act. While my aim is not to make my audiences laugh, although comedy is one of the best ways to bring an audience along with you, I know that feeling of "dying." It's when the impact and power of what I am trying to say doesn't reach their educational hearts and minds.

As soon as I came out with the hear-speak-read statement there was an audible "aha" moment. Teachers strive to re-create the "aha" moment in their students as often as possible, the moment when the penny drops and all the different concepts you have been teaching them fall into place. As I mentioned earlier, I call it a cognitive click and I still have two drawings my daughter did side by side on my office wall that illustrate the concept beautifully. Done on consecutive days, the first is a typical abstract drawing with lots of violent lines and colors filling the entire page, the same drawing she had done since she could hold a pencil. The other drawing, from the very next day, was the shape of a face with multiple eyes and mouths, but unmistakably human. Sometime during the night her brain had experienced a significant cognitive click.

I had somehow made it to the magical land of cognitive click with these school leaders. To them this was no longer a seemingly unhelpful presentation about the often extracurricular or marginalized area of music education, this was now a presentation about one of the fundamental aspects of school education: learning to read.

In this chapter I want you to experience your own cognitive click by bringing together several new models on how music and language development are intimately connected. These will build on the model I first introduced in Chapter 5, the phonological loop, which is the overlapping neural network that links written music and written language. Here is a shortened version to jog your memory:

1. See the symbol in music or language.
2. Hear the brain recording of the sound of the symbol.
3. Tell your body how to make the sound.
4. Make the sound.
5. Check the sound you made with your ears. Correct or reinforce the sound in your brain recording.
6. Repeat.

The researcher's hunch

Most research starts with a question and a hunch. It is often an observed phenomenon, something like avocados seem to ripen faster when you put a banana next to them. A researcher asks questions like: "Why does that happen?" and "Is it really happening or is it a coincidence?" The thing about a hunch, and about the way humans like to search out evidence that proves our hunch was right in the first place, is that researchers want to look into their hunch without bias and be as objective as possible. This leads to the principle of the scientific method. The scientific method is important, although we should also understand it has its limitations when it comes to human development because, let's face it, we are all variations from the "typical."

The first hunch I want to share is the phenomenon that children who can keep a steady musical beat seem to be able to progress in their reading in a standard, typical way. By steady I mean being able to tap a drum or clap their hands consistently to a given beat, not complex rhythms that you might see played on a drum kit. Questions a researcher could ask about this hunch might include: "Do these two skills have a connection or are they just parallel developmental skills around the same age?" or "If these skills are somehow connected or overlapping, why and how could that happen?" And there's always the chicken-and-egg question: "Which came first, beat-keeping or reading?"

In a 2018 study led by Professor Silvia Bonacina, the research team designed an experiment on the back of a large body of other research into elements of this hunch to try to answer these questions. Suspecting there might be multiple networks and functions at play in the brain, they did two tests with sixty-four children between five and seven years old.

The first test was a speech-in-noise test that looked at how accurately a young child's brain processed speech sounds when there was a lot of other noise, in this case, multiple female voices speaking nonsense. I got to watch this test in action with a three-year-old girl as well as doing it myself, and the results amazed me. Both our brains were working hard to dissect the sound but neither of us was aware of the cognitive hard work because we were being distracted by a movie that was playing with no sound. The test was conducted entirely without our conscious knowledge.

The second test was about beat-keeping with two conditions. The first was hearing the beat only and tapping along, and the second condition added a visual element with a dot flashing up on a screen every time there was meant to be a beat.

What did the researchers find? Well, children who had high levels of speech-in-noise processing and could maintain a beat with both just the sound and with the visual feedback as well had higher processing abilities in the areas of the brain that are used for reading. So the hunch was correct, but now they needed to get a picture of the why and how.

The research team put forward a model that brought the two main tests (there were multiple subtests as well) together. In a nutshell, when the children were using just their ears, in the speech-in-noise test and the beat-keeping with only sound, their auditory processing was very stable and consistent. What this means is that the auditory processing kept working throughout the test and didn't drop out and then kick back in—the auditory processing was "phase locked" into the pattern of the sound quickly and maintained that lock, and the start, middle, and end of the sounds (called the envelope) were also consistently accurate. You might compare it to being switched on and engaged throughout an entire meeting rather than your mind drifting off and missing details.

Exciting things happened when they added in the visual element to the beat. Think about it like getting to the next level of a video game where you have to take the skills you were using before and add another element. In this test, the researchers who had already engaged the auditory and motor cortices to follow and make the sound were now adding in a brain part, the eyes.

What they found when the visual element was added was that a group of skills related to literacy kicked into gear. When the children's ears, body, and eyes were working on maintaining a steady beat, other networks and functions such as phonological (language sounds) memory and awareness, basic reading skills, and morphology syntax (form and study of words) were also at work. They also found that the processing speed improved, and this reflected work presented in another model in the field that we will look at next.

Bonacina and her team put forward a possible why and how for their hunch:

> Music learning, specifically the act of keeping a steady beat, helps create auditory processing stability in a child's brain. That stability supports their ability to extend that auditory processing from a beat to speech and language. This starts with phonological memory and awareness and helps them play with the structures and forms of words easily. Once they start to see a visual element connected with the beat a child's brain connects the auditory sound they are making with the drum to a symbol they see with their eyes. This completes the circuit of symbol to sound and means a child's brain is wired for the next challenge to these networks, the ability to read.

Faster, higher, stronger

This phrase is the Olympic motto, but I often think of it when imagining what is happening inside a child's brain as they learn to read. They are trying to read faster, and everyone knows that stilted reading tone when children are first trying to read. They are trying to heighten their vocabulary and understanding of how words and sentences work, and mix some speech prosody or emotion in there, too. And they are trying to strengthen their neural encoding for speech, which is like trying to turn a grassy walking path into a highway in their brains.

The reason we know that these faster, higher, stronger neural characteristics are part of learning to read is that researchers have compared typical and poor readers to see if they can spot the differences. In a cross-sectional study (one that brings together the findings of lots of other studies) published in 2013, Dr. Adam Tierney and Professor Nina Kraus put forward a biological basis for reading problems. They found that children with poor reading have slower neural timing, weaker speech encoding, and lower levels of distinction between speech sounds.

Another reason why schools in disadvantaged areas are studied so extensively is that when music programs are implemented in these schools, students' reading levels seem to improve significantly and rapidly. Reading skills, along with many other skills including executive function and social skills, are often found to be lower in children from disadvantaged backgrounds. Here is another researcher hunch: when music programs are introduced to schools—and it should be noted that the music programs are often intensive, instrumental-based programs that include high-stakes performances at professional venues or on TV—reading levels and scores seem to improve.

The two studies I have highlighted (and if you delve further and read these studies you will see they refer to hundreds of studies that they have based their hunches upon) begin to explain the why and the how. Through environmental and/or genetic factors, children who struggle to read have neural processes that we can now identify as contributing to that struggle. We have also been able to identify a less obvious but possibly highly effective activity that can assist those neural responses to become faster, higher, and stronger.

What about the when?

I am hoping you have had a cognitive click by now, but if you haven't, don't worry, it will happen when you least expect it! Let's get back to those school leaders. The research I have just talked about tends to center around the time when children are typically learning how to read. The saying goes: "For the first two years of school, students are learning how to read; after that, they are reading to learn." For the majority of students this may be true, but the saying itself concerns me, because what happens if a student doesn't get reading under their belt in the first two years? What is learning and school like for them?

This is where research into music and the brain has the most potential to change the lives of thousands of students, and by extension thousands of families and, in the end, the entire social and economic fabric of a country. In the United States, research has suggested that national and state planners could make infrastructure decisions about how many places they will need prisons in ten or fifteen years based on the current rates of literacy in seven-year-old students.[2] The ability to read and, by extension, to learn in our current education system is one of the most powerful early steps in a child's development and future success.

The school leaders, many of them leading junior high or high schools, may wrongly perceive that this research is not going to help their teenage students. For the most part, the aspect I have focused on in this chapter, reading and music, may not be as readily applicable to teenagers as other impacts such as improving impulse control and attention. Those benefits will be explored later in the book but for now, let's focus on the school leaders in our elementary schools. Before I go on, while you might automatically ascribe a school leader to be a principal or head teacher, I see school leaders as those who guide the direction of a school. This encompasses board members, parents, policymakers, leadership teams, and specific curriculum and extracurriculum leaders. We all "contribute a verse" to each student's education, to quote Walt Whitman and *Dead Poets Society*.

As mentioned in Chapter 5, we are beginning to see from the research that music and reading may well be complementary learning activities, but we are only just beginning to understand the how and why. Part of the how and why introduces the concept of sensitivity periods for brain development, meaning a time when the brain is more sensitive to learning.

---

2   Hernandez, D.J. (2011). "Double jeopardy: How third-grade reading skills and poverty influence high school graduation," Annie E. Casey Foundation.

In 2013, Dr. Erin White led a team of researchers looking specifically at this idea as it related to music training and language development. What they found was that there is a sensitivity for motor and auditory development between the ages of six and eight years that both music and language learning rely on. The researchers also looked at the language processing of adults who started learning music before and after the age of seven. They found that there was a notable difference, with those who started learning before the age of seven having better language processing. Interestingly, there also seemed to be a divide between those who learned music for one to five years and those who learned for six to eleven years, with more years of learning relating to better language processing.

There are multiple factors to consider here, of course, but a growing body of research points to an essential and foundational sensitivity period for brain development between the ages of birth and seven. Ignatius Loyola (1491–1556) may have been one of the earliest neuroscientists when he said: "Give me a child until he is seven and I will show you the man." This is the period of development when, as parents, we observe our children learning how to move with intention, speak, read, understand numbers, dance, feed themselves. If I had to keep listing all the things they learn this would be a very long chapter; however, it could be said that between birth and age seven a child becomes somewhat self-sufficient. It is also clear that during this time children get the first opportunity to develop their personality, preferences, beliefs, and dislikes that, in most cases, they will carry into adulthood.

Another factor to take into account is a child's first language. Again, an expanding area of research is the difference in auditory process (and therefore musical and language development) across different types of languages, specifically tonal languages such as Mandarin or Cantonese and stress languages such as English. White and her team suggested that some of the transferable processing in the brain may be different depending on what a child's first language is. The study suggests that tonal languages, which have higher levels of melodic variation, might assist with musical development. Indeed, research has found that children who learn Mandarin or Cantonese as their first language can have significantly higher levels of auditory processing, which may make learning music feel easier. Conversely, English, which is a stress language with far less melodic variation, may not assist auditory development in the same way.

Beat the drum

Considering our children need both auditory and language processing to develop effectively, this research may point to the fact that children speaking English as their first language need music learning more, not less, than we may have previously thought. In a world that is only getting noisier, where our auditory processing systems are being asked to work harder and more consistently than they were designed to, and where we have fragmented the education experience for our children, we need research to help us move to the next stage.

If you are a school leader, in the broadest sense of the term, how is your school viewing music at this particular time in the reading development of your students? Do teachers know about the research and do they understand its implications?

As a parent what would you like to see in terms of music learning for your child and what are they receiving currently?

Here is a place to start: How can we use music learning as a complementary and priming activity for learning to read? Let's ensure that every child gets their language and cognitive click in their first two years of school.

Further reading

Bonacina, S., Krizman, J., White-Schwoch, T. & Kraus, N. (2018) "Clapping in time parallels literacy and calls upon overlapping neural mechanisms in early readers," *Annals of the New York Academy of Sciences*, 1423(1), pp. 338–48.

Kraus, N. & Slater, J. (2015) "Music and language: Relations and disconnections," *Handbook of Clinical Neurology*, 129, pp. 207–22.

Tierney, A. & Kraus, N. (2013) "Music training for the development of reading skills," *Progress in Brain Research*, 207, pp. 209–41.

White, E.J., Hutka, S.A., Williams, L.J. & Moreno, S. (2013) "Learning, neural plasticity and sensitive periods: Implications for language acquisition, music training and transfer across the lifespan," *Frontiers in Systems Neuroscience*, 7, p. 90.

8

CONNECTING THE DOTS

Why keeping a beat is vital for reading

"Okay, let's start by keeping a steady beat."
Clap-clap-clap-clap.
"Now let's walk on the spot to the beat at the same time."
Stomp-stomp-stomp-stomp.
"Very good! Now let's try singing 'Twinkle, Twinkle, Little Star' at the same time."
Sing-sing-sing-sing.
"And stop! Do you think your first grade class could do the same activity in the same time frame?"
Blank.
"No! Maybe? I have to work really hard to do that, and I'm a teacher!"

This is an activity I often do with knowledgeable, confident, highly educated elementary school teachers. Doing three things at once—getting their brains to follow instructions to three different parts of their body with three different actions—is cognitively taxing. But it is the best and quickest way I have found to remind professional educators that learning in the way that their young students are is a workout for the brain, every single day. And the hardest part of this exercise is not necessarily the different actions but the act of keeping a musical beat.

In this chapter I want to connect the dots between two seemingly separate learning areas, music and reading. As we have already explored, these are actually complementary and overlapping processes inside every child's brain.

Once upon a time in a school

By now you will have worked out that I am fond of a story to illustrate a point. Humans remember stories far more readily than they remember facts and, more importantly, they learn *through* story. Here is one that for me illustrates the power of beat-keeping to help us understand where children between the ages of five and seven are in their brain development.

In this story I am working with a group of elementary educators, about twenty-eight of them, in a beautiful and happy school in a challenging socioeconomic area. I will mention only a few things about this type of school. First, the teachers and school deal with a typical and long list of learning and behavioral issues that come from living in disadvantaged, challenging circumstances, underserved populations—whichever term works for you. I could list the issues but would rather put you in the lives of some of their students and see what you make of it.

You arrive at school having fled the house due to another argument. You didn't get dinner last night or breakfast this morning and you are responsible, at the age of seven, for walking your six- and four-year-old siblings to school with the lunches you made for them. Now jump in and learn your times tables and pay attention all day long before going back home to uncertainty and sometimes chaos.

There are as many different stories as there are children, but what these teachers manage is not just what happens in school hours but the whole child, who in this case is impacted by an unstable home life. These are the teachers I enjoy working with the most. Due to their challenging professional environments and clientele (both students and parents), they are open to whatever works in education. They are highly attuned to different needs on any given day for each of their students. This is not to say that teachers in less challenging areas are not like that, but it has been my observation that when faced with higher levels of fluctuation and variety in your students, you become an intuitive, highly observational, innovative educator pretty quickly. When I have the great privilege of watching them teach, I feel like I am observing true artists at work.

Imagine these twenty-eight teachers have just finished an active music activity, clapping, stomping, singing, and now they want to show me a video of their final assembly last term where one of the classes performed a musical activity I had taught them. They are buzzing with excitement.

Play the recording. Twenty students (six and seven years old) are seated on the edge of a low stage. They are clapping to the beat, alternating between their knees and hands, and saying a rhyme in time with the beat. I'm beaming because the activity has all the right music education elements: clapping in time, saying a rhyme, and paying attention to the teacher for more than ten seconds. The teachers are beaming because they have taken on the challenge to implement music activities into their everyday teaching, and to get to the point of a performance is a huge milestone. And then I watch more closely.

After the short video has finished, I congratulate the teachers. But instead of moving on to my next slide I take what is called a teachable moment, when you go off script with the vague intuition that there might be more here that is powerful and important to learn.

"Can I try something?" I ask the assembled teacher–hive mind.

"Sure," they openly reply.

"This boy on the end. I can see that he is really struggling to keep up with the beat, almost like he has to see it first to then tell his brain to move his arms to clap. Also, his eye contact attention moves every three beats. I wonder if he is counting in his head, or if his attention span needs a reset every few moments. He seems taller than the other students, too, and his torso looks a little slumped and not supporting him. Is he quite behind in his reading and does he seem to get tired easily? Can he catch a ball?"

I don't know these children at all, and they don't know me. I am purely going on what I can see from a musical and sensory-motor perspective and checking if it lines up with the research I have read.

There is a pause and then boom—it was like I had a crystal ball in front of me.

"How did you know that?"

"He has never caught a ball in his life!"

"I'm his teacher and I just don't know what to do with him. I can't figure out where the block to his learning is."

These external indicators of learning difficulties are not new; any experienced early childhood educator could spot them at twenty paces. But what interests me in this field of research is finding an answer to why this boy is struggling with keeping a steady beat and with his reading and what is happening inside his brain. My answer directed to his class teacher and the greater assembly of teachers was: "Well, the research says ..."

It's all in the timing

"Clapping in time represents an activity that almost everyone experiences since childhood. It requires global coordination and interaction between motor and sensory systems and a fine temporal ability to control the entire movement as to be on time." It is the very fact that we as adults have been doing it since childhood that has given us so much practice at this global coordination. Additionally, we are beginning to discover the other neural connections and subsequent skills that we build on top of this global coordination throughout our lives. We also know that keeping a steady beat can be used as a thera-peutic activity after a traumatic brain injury to reengage that global coordination after it has been damaged.

The statement above comes from a research paper entitled "Clapping in time parallels literacy and calls upon overlapping neural mechanisms in early readers" led by Silvia Bonacina, so you can probably tell where this section is going. The words that pop out for me in it are *coordination*, *interaction*, and *fine*. The little boy on the end of the stage has the coordination to clap his hands but not the interaction between his auditory and motor systems to coordinate the action with the beat. What's more, he doesn't have the fine temporal (or time-based) ability to get back onto the beat during the activity. He remains off the beat the entire way through the rhyme. This lack of direct timing and the inability to change the timing all contribute to the lack of a neural foundation upon which to then build the skill of reading. Reading and music use many of the same areas of the brain and also overlapping networks, so it stands to reason that one will influence the other. Interestingly, music is easier, and more innate, to the brain than reading. It may seem strange but parts of our brain are different ages. What I mean is that the human brain has evolved over time and some parts are newer, such as our prefrontal cortex where our executive function lives, and some parts are older, meaning they were some of the first to develop. Music, and the elements of music such as melody and rhythm, may be one of the older and more established parts.

"Has his reading changed at all this term as you have been practicing this activity for the assembly?" I ask, fingers crossed.

His teacher speaks up at this point. "Actually, yes—not as much progress as the other students, but for him the improvement has been noticeable."

Another teacher adds, "You know, he came into the library the other day and actually pulled out a book. I had never seen him in there before."

That brings a big smile to my face. I like this game. Let's try another student on the video!

Knowing what is coming next

Prediction—a child's ability to predict what is coming next—is being revealed in the research as a specific difference in brain development that then affects learning. Prediction is influenced by expectation, that is, what we expect will come next. Music is actually a really good way to explain prediction and expectation, and it turns out it is a great tool to study it. When we are listening to a pop song we are quickly set up with the patterns of the music, maybe a catchy bass line. The structure of that bass line will be conforming to an expected pattern that most people will find satisfying. Interestingly, when that bass line is changed a little, but not too much, our brains find that violation of expectation satisfying as well. Our brains use prediction all the time, from predicting when we can safely step off a curb after a car has passed by to complex cognitive predictions involving multiple possibilities and factors.

"The girl fifth from the end—if we turn the sound off, can you see that her clap comes when it should but the actions before and after are very jerky and don't flow in that kind of circular motion of a continuous clap? It is almost as though she thinks she is going to miss the beat, so she races to get it. I am wondering, does she struggle with integrating new words? What I mean is that she really likes patterns of words she knows, but when an unexpected word is used, does she find it hard to integrate that into her vocab?"

"Yes!"

"Does she need a lot of repetition, not just in reading but everything?"

"She does love routine, and I think she gets anxious when the routine changes, like the last day of school and this assembly, actually."

I nod. "I'm worried she won't be able to adjust to change, and life is all about change."

Enter the jazz musician

Jazz musicians have fascinating brains. They are truly wired differently, and as a music educator I sometimes get stuck in an existential question: Does learning jazz, and specifically improvisation, develop a student's brain in this particular way, or are those students who already have that specific wiring drawn to the genre of jazz and the experience of improvisation? Once again, I think it is a bit of both.

Jazz musicians have been studied for many reasons, but one in particular is a faster and better capacity to predict what comes next. Musically speaking, improvisation is structured around a musical key and melodic concepts but it is also seemingly unstructured when it comes to responding to or predicting where another player's melody is going to go. Studies have looked at the flow or meditative state of the brain during improvisation because jazz musicians access that flow state regularly.

It is interesting that the ability to predict what is coming next, like the next note in a song or the next word on a page, develops in conjunction with the ability to identify when the predicted note or word doesn't appear. But we are now beginning to learn that to develop one is to develop the other. Music is filled with patterns, just like mathematics is filled with patterns, and music learning in the early years of school is filled with pattern prediction, violation, and creation. The patterns are made with sound and are often connected directly with symbols as well. Music activities using beat and patterns of sound (called rhythms) are delivered through the body at this early stage of education. This means they engage all those areas from the statement on clapping in time I quoted earlier, the global coordination and integration with fine temporal activity. For the little girl struggling with change, musical activities might be just the thing to connect the predictive dots in her brain.

Hiding my crossed fingers behind my back, because doing the crystal ball trick twice was just too much to hope for, I ask her class teacher the same question: Has her reading changed since you started doing this musical activity?

"A bit, but the biggest change I have seen is she is now using her finger to follow the words. I said to her just this morning, make sure your finger keeps up with the words you are saying. I think she will get it, but I had no idea it came from this idea of predicting what comes next."

Bingo!

Now we are rolling

I think about going for the trifecta of fortune-telling but I don't get the chance. These teachers are so excited by this new technique that helps them view their students differently and the potential to approach old problems in a new way that they take over.

"Okay, I want to give it a try," says one particularly astute teacher. "See the girl in the middle, the one who is sitting up really straight? Now, she is a middle-of-the-road learner, progressing at the standard pace with her reading. But she's a chatterer, she could talk underwater." (Much laughter from the other teachers.) "I think she has the potential to progress far more quickly, but this sort of distraction in the form of talking consistently gets in the way. So since we started this activity she has done exactly that, she has knuckled down and gets less distracted, and she has moved up four levels in one term when she might only move up one or two in a regular term. Could it be something to do with the music, and if so, what is it?"

I take a moment and try to repress my happy dance.

A lot of things get in the way of learning. Learning takes consistent attention, repetition, and engagement. It pulls on our sensory, memory, and cognitive systems and can be disrupted by anything from an empty stomach to a friend who isn't at school today. Staying on task is one of the hardest challenges we have in our modern lives and you can tell by the large number of books about productivity that this is a struggle for us all.

Keeping on track is both a nature and nurture skill. Researchers believe we may be born with predispositions for focusing as well as learning focusing skills and literally what it looks like from role models, specifically parents, teachers, and peers. One of the many executive functions that we use to both get us back on track and keep us there is inhibitory control. This is the ability to both stop getting distracted by anything outside the task as well as pulling ourselves back into the task when we have wandered off the track.

As simple as it sounds, keeping a beat is almost the perfect training ground for inhibitory control. Music is learned in groups, and groups provide positive peer pressure, meaning you keep going because you are in a group. Think of group fitness versus individual fitness: Which one requires more effort to remain on task? Keeping a beat is a discipline but after a while it becomes automatic, meaning we don't use too much of our conscious brain to continue doing it.

If music and reading share overlapping neural networks, then it may be possible that training an executive function skill through music, like inhibitory control, may transfer in some way to reading. It's probably safe to say that the ability to maintain attention when we are learning to read will likely lead to improved reading progression, but researchers are also looking more deeply into the specific reading areas that may be affected. For example, dyslexia has been a hot topic of research, as this disorder presents as a reading issue, but may have underlying factors that come from auditory processing and attention deficits. We are at the beginning of this exciting field of research but keeping the beat, leading to greater brain synchronization, and learning a discipline that includes our whole bodies can result in greater focus and improve learning.

The rhythm of life

It is fascinating to me that beat and rhythm are fundamental to our internal and external lives. The research continues to make me look at beat and rhythm in all of their forms: from a physiotherapist who can see the rhythm in someone's gait and diagnose which muscle is giving them trouble, to a psychologist who understands rhythms within a person's psyche, to a stock trader who talks in hushed tones about feeling the rhythm of the market. Beat and rhythm are in everything if we look and listen. In the examples I have shared in this chapter the beat was in, or not in, the students' bodies, and it helped us understand where their brains may be working effectively and where they may not be. We have a tendency in life and in schools to remain task-focused and not see and hear the broader beats and rhythms for ourselves, our children, and our students. Use this concept to look anew and afresh at our children and our students, because the closer we can pinpoint the deeper cognitive issues that may be present, the better we can be at helping them to thrive and read.

Further reading

Bonacina, S., Krizman, J., White-Schwoch, T. & Kraus, N. (2018) "Clapping in time parallels literacy and calls upon overlapping neural mechanisms in early readers," *Annals of the New York Academy of Sciences*, 1423(1), pp. 338–48.

Corrigall, K.A. & Trainor, L.J. (2011) "Associations between length of music training and reading skills in children," *Music Perception: An Interdisciplinary Journal*, 29(2), pp. 147–55.

Dittinger, E., Barbaroux, M., D'Imperio, M., Jäncke, L., Elmer, S. & Besson, M. (2016) "Professional music training and novel word learning: From faster semantic encoding to longer-lasting word representations," *Journal of Cognitive Neuroscience*, 28(10), pp. 1584–602.

Hornickel, J. & Kraus, N. (2013) "Unstable representation of sound: A biological marker of dyslexia," *Journal of Neuroscience*, 33(8), pp. 3500–504.

Janus, M., Lee, Y., Moreno, S. & Bialystok, E. (2016) "Effects of short-term music and second-language training on executive control," *Journal of Experimental Child Psychology*, 144, pp. 84–97.

Steinbrink, C., Knigge, J., Mannhaupt, G., Sallat, S. & Werkle, A. (2019) "Are temporal and tonal musical skills related to phonological awareness and literacy skills?: Evidence from two cross-sectional studies with children from different age groups," *Frontiers in Psychology*, 10, p. 805.

Vuust, P., Ostergaard, L., Pallesen, K.J., Bailey, C. & Roepstorff, A. (2009) "Predictive coding of music–brain responses to rhythmic incongruity," *Cortex*, 45(1), pp. 80–92.

9

THE SOCIAL SOUP

How music helps your child learn control

Have you said any of these lines more than once?

"Don't grab, just wait your turn."

"Let her finish her turn and then you can have a go."

"It's time to get off the swing and give the other children a go."

The concept of sharing and, more importantly, waiting our turn, is a tough one for young children. We try to teach it in all sorts of ways. These can be explicit and direct—I still remember when my very young daughter piped up one day after school and announced, "Sharing is caring, Mum." (Interestingly, when I asked her what sharing looked like she said it was being nice to other people, nothing about actually sharing or waiting your turn.)

As parents and educators we also teach this difficult idea to young ones by role modeling. We show intense attention in our bodies and faces during a speech in assembly, even when we might not be that interested. We might share our food at the dinner table in an overtly equitable manner or insist on waiting until everyone is seated and ready before beginning to eat. This last one can be tough with a starving six-year-old, but the ideas and the actions of sharing and waiting are fundamental skills in behavior development for our children.

Both of these skills require control, specifically what is known as impulse or inhibitory control. In education we tend to use the term impulse control and in brain research it's usually inhibitory control, and by definition they are a little different, but for the purposes of this chapter I will use the term *inhibitory control* when referring to the scientific research. I personally like to call it "I want to eat the chocolate cake, I shouldn't eat the chocolate cake" control! In this chapter we will first look at why this skill is so hard for young children to learn and then explore what music learning has to do with the decision to eat chocolate cake.

The marshmallow experiment

I know I am presenting you with an array of sweet temptations in the very first pages of this chapter, so if you feel the need to go and grab your favorite treat right now, I totally understand. Sweet treats are used in inhibitory control experiments because humans tend to crave sugar, and that craving tests our cognitive control processes. We will get back to music learning soon, but first let me describe possibly the most quoted inhibitory control experiment with children: the Stanford marshmallow experiment.

In a nutshell, this is a series of experiments exploring when children begin to develop the ability to delay gratification, which basically means being able to wait a short period of time in order to get a reward for waiting. While there are many versions of the test, which have now been conducted since the early 1970s, here is the general gist.

A child between the ages of four and five years is asked to sit down at a table where there is a single marshmallow on a plate. The researcher explains they can eat the marshmallow now if they like, but if they wait for fifteen minutes without eating it then they can have two marshmallows. The researcher then leaves the room and observes their behavior. Children had some very interesting reactions to this information, with some of the reported behaviors ranging from covering their eyes so they couldn't see the marshmallow to stroking the marshmallow like it was a tiny stuffed animal.

In the Stanford study published in 1972, a small group of children ate the marshmallow straightaway, two thirds of the children attempted to delay their gratification but failed to last the distance, and about one third of the children made it past the fifteen-minute mark and received an additional marshmallow. Where this experiment became very interesting was when these children were followed up with ten years later. The researchers wondered if the ability to delay gratification between the age of four and five had any connection with inhibitory control in later life, specifically the ability to cope with frustration and stress in adolescence. The researchers found there were statistically consistent correlations between the one third of young children who successfully delayed gratification and their ability to manage the pressures of adolescence.

What does statistically consistent correlation mean? It essentially means that while we can't say directly that success in the marshmallow test leads to success in managing frustration and stress in adolescence, the researchers felt pretty sure they were connected. Replication of this experiment has continued over the years, and a more recent 2018 study also found a correlation, but the statistical significance of it was far less. Yay for science—we have so much more to learn about inhibitory control, all inspired by the humble marshmallow.

Inhibitory control in the twenty-first century

From the start of this century we have seen a rapid expansion of technology that has made our lives easier, arguably more connected, and definitely more immediate. Children are using technology to access all the knowledge of the world, both fake and fact, and education theory and practice is scrambling to figure out what on earth twenty-first-century skills might be.

Another by-product of these advances is that children can now watch TV, play games and access information whenever they want. No more waiting for your favorite show to come on every Thursday at 7 p.m., just switch on your favorite streaming service. No more waiting for friends to come over after school to play your favorite game, just jump online and make your next move digitally—who needs to be in the same house, suburb, or even country.

These technological developments are amazing. But with everything at our fingertips, how do our children learn the great art of waiting? This is where the study of music learning, among many other activities, comes into play.

The musician's brain

Research into music learning and brain development started with adults, mostly professional musicians. This was because adult musicians were found to have brains that were wired very differently and highly effectively, and researchers wanted to learn more about what might have caused this. Again, it is a classic chicken-and-egg conundrum: Did smart people take up and excel at music learning or did music learning change how their brains developed?

One of the very first areas that researchers found to be highly developed in adult musicians was their inhibitory control. Anyone who has attempted to learn a musical instrument, including voice, will know how much discipline goes into that learning. You have to get it wrong again and again and again before you get it right. Even then, you might get it right in your Tuesday afternoon practice but not be able to play it in your Wednesday morning rehearsal. It is infuriating and, as it turns out, wonderfully instructive for our brains.

One way the higher level of inhibitory control kicks in during the process of learning music is during the more frustrating and stressful parts of practice, when you just can't seem to get it. Strong inhibitory control can stop us from giving up on the note we are trying to reach or the run of notes we are trying to put together, and we might call this persistence or sticking to it. Whatever term you use, the inhibitory control is stopping a musician from giving up, walking away, or throwing in the towel. The little doses of frustration and stress that adult musicians have learned through many years of exposure help them to recognize, manage, and conquer future frustration. It is like exposure therapy for frustration.

But is inhibitory control inherited or learned? As I find myself saying so often in any area of development, it is most likely both. Research into genetics and individual personalities has shown that both factors play a significant role in the level of inhibitory control an individual possesses, and these levels could be the deciding factor when it comes to musically trained children who appear to find the activity easy or have what we might identify as talent. Another group of researchers have sought to test when and how inhibitory control might be affected by experience, specifically the experience of learning music, which appears to have some unique qualities.

It's not your turn to play yet

Once identifying that inhibitory control was high in adult musicians, researchers started working backward, or developmentally. When did children begin to show behavioral choices that indicated inhibitory control, and if they learned music did that activity have any impact on those inhibitory skills?

In the original marshmallow experiment with the children between four and five years of age, one third of the experiment group successfully delayed gratification, so this developmental stage could be the start of inhibitory control development. Three research teams designed studies with randomized groups of children who commenced their music learning between six and seven years,[3] eight and ten years,[4] and nine and twelve years.[5] These studies looked at slightly different areas of inhibitory control, often in conjunction with other executive function skills such as attention, intelligence, or planning. All three studies found marked improvement with inhibitory control which, when the results were controlled for many other factors, could be attributed to the music learning experience.

Indeed, the 2011 study led by Dr. Franziska Degé in Germany proposed the concept that the improved inhibitory control in musically trained children was the mediating factor between musically trained children and higher levels of intelligence. Put another way, the ability to delay gratification and wait your turn was the skill children needed to then go on to develop their overall intelligence. A study published as recently as 2018 led by Dr. Artur Jaschke in the Netherlands supported this connection after following 147 six-year-old children for more than two years of music learning. The researchers concluded that through their music learning, the children in the study improved their executive function abilities of inhibition

3   Habibi, A., Damasio, A., Ilari, B., Elliott Sachs, M. & Damasio, H. (2018) "Music training and child development: A review of recent findings from a longitudinal study," *Annals of the New York Academy of Sciences*, 1423(1), pp. 73–81.

4   Fasano, M.C., Semeraro, C., Cassibba, R., Kringelbach, M.L., Monacis, L., de Palo, V., Vuust, P. & Brattico, E. (2019) "Short-term orchestral music training modulates hyperactivity and inhibitory control in school-age children: A longitudinal behavioral study," *Frontiers in Psychology*, 10, p. 750.

5   Degé, F., Kubicek, C. & Schwarzer, G. (2011) "Music lessons and intelligence: A relation mediated by executive functions," *Music Perception: An Interdisciplinary Journal*, 29(2), pp. 195–201.

and planning, and these improvements acted as the transfer mechanism across to their improved academic achievement.[6]

When is it my turn to clap?

This research is all well and good, but for parents and especially school leaders who are interested in why music learning improves inhibitory control, and how in turn this improves sharing and waiting your turn, let me paint a picture for you.

It is the lesson before lunch and twenty-five seven-year-olds are hungry. It's a hot summer day and the teacher is ready for a break, too. There are three precious minutes left of learning time—what can we do with that? We can fine-tune the students' still-developing motor control, instill a sense of deep-seated teamwork and respect for each other, challenge their auditory processing networks to be aligned with their visual networks, and give them a daily dose of inhibitory control development, that's what. Line up at the door, everyone!

The challenge is simple. The teacher gives the class a beat. To be clear, a beat is a repeated sound that is equidistantly apart, or basically the time that your foot keeps when you are listening to a song you like. To gain exit from the classroom and entry to the delightful freedom that is lunchtime, the class needs to complete one simple task: pass the steady beat down the line without missing a clap or a student. While parents might be thinking, "Well, how hard can that be," any teachers reading this are probably laughing right now: hungry,

---

6   Jaschke, A.C., Honing, H. & Scherder, E.J. (2018) "Longitudinal analysis of music education on executive functions in primary school children," *Frontiers in Neuroscience*, 12, p. 103.

hot seven-year-old children can't focus or work as a team that deliberately.

Why is it hard? Because each person in the line needs to activate their inhibitory control, supercharge their visual, auditory, and motor cortices to connect in a single moment and engage their prediction skills to execute their role in the larger challenge. And all the time this cognitive and emotional connectivity and performance is being compromised by hunger, mental tiredness, temperature, and boredom. Overlay this with the pressure that if you drop your attention for a moment you let the whole team down and have to start again. Not a recipe for social acceptance and support in seven-year-olds, or possibly even grown-ups.

This is a real-life experience, so let me tell you what happened with those hot, hungry seven-year-old children. On attempt number one they got one third of the way down the line before a student who wasn't great at paying attention jumped in too soon. On attempt number two they got halfway down the line, but it was shaky around the inattentive student. Another student asked him, "If you stood on the other side of me next time and I clapped loudly, would that help?" They switched spots and on attempt number three they got past him to the fourth student from the end, but the pressure was too much and she hesitated. Oh, the groans and faceplants! The clock was ticking and lunch was fast approaching. The girl fourth from the end looked worried: What if she failed again and let everyone down?

Suddenly a collective decision to start bobbing heads and bouncing bodies to the beat engulfed the entire class, like rock music was playing in the background. They started again, determined to succeed, and the bell sounded three quarters of the way down the line. A student is startled by the bell and they lose the beat. "Nooooooooooo!" But then the magic really happens. The student at the start of the line pipes up and says, "We can do this, one last try." They could have given up, they could have said we can't do it but who cares, it's lunchtime. But they didn't. On attempt number five they got it, shaky clappers and nervous team players all. They connected their brains and activated their inhibitory control while all the time completing a task as a team for the sake of it, not for a mark or any other reward other than a sense of achievement.

The power of sharing and waiting

Infused into many music learning experiences at this age are the concepts of waiting your turn and sharing an instrument or a rhythm. The power of these experiences at this age is they are taught by microdosing, short and small experiences of waiting for the right time to clap to a beat or rhythm or swap from playing a castanet to a tambourine. This becomes even more pronounced when children learn an instrument in an ensemble setting. If you have ten six-year-olds learning violin, they can't all be playing at the same time. They need to have strict procedures about when they lift up their instrument, when they play, when they listen to the sound they are making and when they pack away. All of these procedures are microdosing of inhibitory control.

Any parent who has the great pleasure of attending an end-of-year concert where the class gets up to sing a song will be able to see inhibitory control in action (or not). Usually there is at least one student who can't stand still, is looking all over the room and possibly poking the student next to them. This is the student who is struggling to get her inhibitory control into full swing. But for the other students who are standing still, attentive and engaged, that is inhibitory control at work. What we might not get to see as parents is how the music rehearsal process has enabled that control to develop. This is done by creating a shared goal of performing well at the concert and each child battling with their own version of inhibitory control, whether that be a fidgety body, memory problems with the song, or wandering attention. This is all done with the implicit and powerful social soup of achieving something together in a moment in time.

Dig out the marshmallows

As adults we get used to waiting, whether it be waiting in queues or waiting for holidays. We have developed that skill. We have also hopefully developed the skills to resist the chocolate cake that we would like to eat but know for whatever reason we probably shouldn't. These are learned skills, ones that we have developed through often tough but instructional experiences. As life gets easier in some ways for our children as they grow older, the need for activities that continue to teach these necessary skills should become more important and better understood.

Researchers have been working to understand why music learning—an experience that is like a wrapped-up Christmas present, developing many executive function skills such as inhibitory control, attention, focus, and planning—seems to be so effective. The answer is that to produce the performance at the end, whether a song at a concert or a challenge game at the end of a lesson, all the higher-order skills of being human have to not only be engaged but integrated within ourselves and with the other students in the musical team.

It has been proposed that inhibitory control is the mechanism that transfers the experience of music learning and performing to academic achievement. The ability of a child to stick at that tricky math question for just a bit longer until he sees the solution, the ability to keep writing that sentence when there is a word he can't spell, and the ability to not run out of the classroom when it is his turn to present a speech are all examples of inhibitory control in action. There is a strong focus on resilience in both schools and professional life, and inhibitory control is a foundation skill in that process.

How do we help our children and our students see the bigger goal of two marshmallows instead of one, and how do we help them experience waiting in a positive way? Parents sometimes tell me, "My daughter isn't enjoying learning her instrument anymore so I am going to let her stop." Instead of just saying, "Oh, that's too bad," I now sometimes say, "Okay, but do you know how much her brain is growing right now?" It might seem an odd response but I think we should shift the idea that music learning must be fun and easy to the idea that music learning is challenging. The moment when that young girl breaks through the frustration and can do it might be a little way away, but it will be more than worth it.

Further reading

Mischel, W., Ebbesen, E.B. & Raskoff Zeiss, A. (1972) "Cognitive and attentional mechanisms in delay of gratification," *Journal of Personality and Social Psychology*, 21(2), p. 204.

Schlaug, G. (2008) "Music, musicians, and brain plasticity," in S. Hallam, I. Cross and M. Thaut (eds), *Oxford Handbook of Music Psychology*, pp. 197–207.

Shoda, Y., Mischel, W. & Peake, P.K. (1990) "Predicting adolescent cognitive and self-regulatory competencies from preschool delay of gratification: Identifying diagnostic conditions," *Developmental Psychology*, 26(6), p. 978.

Watts, T.W., Duncan, G.J. & Quan, H. (2018) "Revisiting the marshmallow test: A conceptual replication investigating links between early delay of gratification and later outcomes," *Psychological Science*, 29(7), pp. 1159–77.

10

SOUND SYMBOLS

Why reading words is a process of reading music

"Anita, can you read the first paragraph out loud for the class, please?"

I froze. It felt like my stomach fell straight through the floor. There was a buzzing in my ears and I suddenly felt hot all over. In elementary school this was my nightmare moment and it happened almost every day. For a nonreader (or, by the age of nine or ten, a reader who could struggle through sentences but found it extremely hard work), reading out loud in class was, to me, an exercise in public shaming.

Looking back on it, I doubt whether my teachers or friends noticed my struggle. My out-loud reading was stilted and I got the stresses in words confused (my all-time low was the whole class laughing when I pronounced "determined" as it looks—"deter-mined," not "di-ter-minned") but it was passable. Inside my head, though, I was working so hard to connect the symbols on the page with the sounds they should have made. Reading a paragraph out loud was like running a cognitive marathon while inside being on an emotional roller coaster.

Fast-forward to reading out loud to my daughter when she was two years old. I didn't have the freeze response because it was just me and her, and she was more excited about the pictures than the story. All I had to do was make it sound exciting and infuse my reading out loud with loads of emotion.

This low-stress reading-out-loud opportunity got me thinking about how my brain might be wired in an "interesting" way for reading out loud and how the research I had been studying might explain how reading out loud felt for me. Reading to my daughter became an unexpected gift, allowing me to observe my own brain in action at the same time as watching her putting the pieces in place for her own language puzzle. In this chapter we will look at the sound-to-symbol process and how the seemingly natural process of learning to read is actually an incredibly complex and monumental cognitive hurdle.

A cognitive task of herculean proportions

Learning to read has a long and highly researched history. It is a fascinating process that involves many cognitive processes and environmental factors at once, and while we now know an incredible amount about the process, every year teachers encounter young children whose struggle to read is a mystery to them. This chapter is not in any way meant to be a new and revolutionary approach to teaching reading, and I deeply respect the many language and literacy experts I have worked with who know far more about this process than I. This is also not an argument for or against the pedagogical concepts of whole language, teaching language as a whole system from which students can understand its parts, versus phonics, teaching language through its smaller sounds first upon which students build an understanding of language as a system of words, sentences, meanings, etc. This chapter will look at one small aspect of the process, the new understandings that have come from studying musically trained children, about the cognitive connection that underpins part of reading—the sound-to-symbol system.

The sound-to-symbol system

This process is what it says on the packet: the brain's ability to look at a symbol and hear the sound that that symbol represents inside their head. It goes both ways: we start with sound-to-symbol, which goes from spoken language to what it looks like on the page. We then progress to symbol-to-sound when we start to read out loud or inside our heads. In reading research the process might be known by other names (such as phonological representation) and has many sub-parts, including phonemic awareness, transitions, and envelopes. These are all parts of language that a young child's brain needs to distinguish between in order to then put them together in spoken and written language. In Chapter 1 we looked at the complex processes involved in separating speech from noise and then dissecting the parts of speech. The sound-to-symbol system builds on these auditory processing steps and, interestingly, can be made more difficult if the auditory processing foundations aren't as solid as they could be.

What does that mean? Well, as one leading researcher summed it up for me, after about an hour of trying to get my head around the complex brain science: if a child cannot hear it, they cannot speak it, and if they cannot speak it, they cannot read it. She was putting together for me a basic framework of typical developmental processes in the human brain that lead to the ability to read independently.

She explained it like this. A child first learns to dissect the sounds of language but does so in the context of all language: they are hearing the parts within the whole. Musically, this was an interesting idea to me as one of the earliest areas investigated in musicians' brains was the ability to process the whole (all the music) and its part (the drum rhythm in relation to the bass part) simultaneously.

"Is language the same type of process?" I asked.

"At its foundational level," she said, "yes."

This step is squarely in the auditory processing network of the brain, but in order to speak children need to engage their motor networks. When young babies make goo and gaa sounds it is this process in action, hearing the language and then trying to wrap their brains and bodies around how to make that sound. The very patient researcher explained that as a child's auditory processing becomes more detailed and distinct, the language sounds they can make are more distinct as well and the connections between the auditory and motor networks become stronger and more complex.

So how are these processes connected to reading? Well, this is where the eyes come in. A symbol on a page of text or on a musical score is a visual representation of a sound. The child will look at this symbol and immediately hear the corresponding sound inside their brain. Incredibly, we now have ways of recording what the brain hears, the step beyond what the ears hear. With current technology it sounds like a scratchier version of the sound we hear, like an old record. This is because along with the original sound there is something called neural noise, extra sound information that our brain needs to process before deciding if that is part of the sound.

Back to the reading process. The eyes take in the symbol. The visual cortex then connects with the auditory processing network. Think of the biggest iTunes library of sounds you can imagine and the eyes are looking for the sound that is tagged with that symbol. If the correct tagged sound is found, the auditory processing network then instructs the motor cortex in how to make that sound. Take again the example of a "th" sound: I see the letters, I hear the sound in my brain, my brain tells me to put my tongue between or just behind my teeth and push just the right amount of air through to make my tongue vibrate against my teeth while I also engage my larynx to make a sound. It is a complex motor movement for a sound that we say and read a lot in English. If you are a bilingual speaker and reader, try thinking of another common and complex motor movement sound that your brain has to visually, then auditorily, and then physically process.

When I put this process into my own words for the researcher, she reaffirms that this is so at its foundational level, meaning a surface understanding of the process. Beneath each process are subprocesses that are so incredibly complex and interconnected that I would need ten more books (and ten more PhDs) to explain them. But for the purposes of this chapter, a top-level process is a good start. Here it is again in a nutshell:

1. The eyes see the symbol.
2. The visual cortex processes the symbol and connects it to a sound in the big iTunes library of language sounds.
3. The auditory processing network connects to the motor cortex and cerebellum and explains the physical movement required to make the language sound.
4. The body makes the language sound.
5. The sound that is made immediately goes back into the ears and the auditory processing network checks with the big iTunes library recording. Does it match the recording? If it does then it is a tick. If it doesn't, do I need to change my recording or try to make the sound again?

Now here is something to blow your mind. Think about the steps you have just read. How fast is that process going?

What does this have to do with music?
The process of reading a musical symbol and producing a musical sound (either with the voice or on an instrument) and reading a language symbol and saying the sound are, at a foundational level, the same within the brain. A musical symbol such as a crotchet, or quarter note, requires the same neural connections as reading words, just like traveling along the same road to get to the grocery store and the pharmacy.

Just because these two processes share a neural network doesn't mean our brains need to learn to read music in order to learn to read language. Many of you will be wondering about the connection, especially if you or your child can read language but you or your child never learned to read music. The research is not pointing to a requirement for music reading before language reading but what it is pointing to is the possible connection between why children who are involved in musical activities show enhanced levels of reading. It is not just the sound-to-symbol system connection, it is the development of the auditory processing networks before and during the herculean cognitive task of learning how to read. They can be seen as complementary cognitive activities and in much of our educational history have been intricately connected by educators in their classrooms.

When things go "wrong"

There is a vast difference between observing the human brain processing written language and coming up with a conceptual model for what is supposed to happen and the lived experience of learning how to read and watching your own child learn to read. There are many other filters this process needs to go through, such as individual personality, emotional responses, and learned behaviors that impact the complex process of learning how to read. This is one example where the study of learning processes is dependent on whether you are looking through the lens of neuroscience or the lens of psychology.

Naively and inadvertently, I stumbled on a philosophical war raging in the field of neuromusical research during my study tour of leading labs around the world. Neuroscience is the shiny new science, a revolutionary way to understand the mysteries of the human brain. It has given the world some absolutely incredible new insights and will continue to do so, but does it follow that just because this is what happens in the brain that it will transfer to how we learn and behave as humans?

This is the question I put to one of the leading researchers in this field, Professor Glenn Schellenberg from the University of Toronto Mississauga. Professor Schellenberg is a psychologist and pioneering researcher in the neuromusical field. His study on music training and IQ is a seminal piece of research and he has held many others in this field to account when it comes to the rigor of their research and the validity of their findings. Professor Schellenberg argues that just because we see the brain function in a certain way doesn't mean that the human attached to it will function in the same way. There are many other factors that require consideration, in particular personality and genetics, and these may prove the causal and correctional claims that music learning enhances brain development null and void. It is a fascinating debate and one that will keep the field of neuromusical research active and rigorous for decades to come.

This tension between conceptual models of reading and the lived experience of learning how to read could be useful to parents and teachers when reading goes "wrong." I use the word "wrong" quite deliberately because if the research tells us anything it is that learning to read is both a neurological and behavioral experience, and there is no single "right" way of doing it.

My "interesting" brain

In theory, my inability to read easily could have been a problem with my auditory processing. If that was the case then in the hear-speak-read process, my teachers and parents should have seen issues with my speech. There should have been some type of indication that I wasn't hearing the speech sounds correctly and therefore couldn't transfer those sounds to the written word. But that wasn't the case. I was, by all accounts, an articulate child with high aural language levels and I was called a chatterbox more than once. But when I went to read, I had to read very slowly and I would get tired very quickly. Without knowing it I developed techniques to cope with this. For example, I would read a paragraph then take a mini break, staring into space for a minute, and then I would read the next paragraph. Two pages took a long time. I also took to summarizing paragraphs as I went, writing one sentence for each so that when I went back to review I wouldn't have to read as much but still got the idea.

There was definitely something going on in my interesting brain, perhaps issues of connectivity or maybe my iTunes library was not tagging the sound correctly. I will never know because part of what was most interesting about my brain is that it rewired itself in order to work effectively. Maybe if we put personality into the mix it's the fact that I have a good dose of stubbornness in me and a need to achieve. Did these psychological factors impact on my neurological wiring and my learning development, and how did music learning impact on that mix?

Dillon's story

Last year I met a young student who was just about to turn ten years old. He had moved around a number of remote schools and hadn't had the best time socially. Dillon was happy to be at a new school and there was something about him that was quite endearing. He was witty, effervescently happy, and jumped in at any new challenge or experience.

Then I saw him read. Or, I should say, try to read. As a pretend reader myself I both saw and felt it all. I was back in my first grade teacher's classroom again and with every word I was forming a view of myself as a learner that was not positive. Was he another "interesting" brain?

I listened to Dillon's speech, which was fast but not well formed. Letters got muddled up sometimes, he would put a "p" where there should have been a "b," the sounds in the middle of a three-syllable word wouldn't be correct and sometimes he sounded like he had a lisp but he didn't. His body movements could be jumpy and ill-timed but it was all rolled up in this very energetic ball of happy.

I met Dillon because he was part of a music program that was being established at the school. He wanted to play saxophone because it was "loud, just like me." He was partnered with another student who had a very steady character, an average student well known for being able to work independently.

As part of his music learning, Dillon was taught how to read music. He became friends with his music partner and a good dose of competition grew between them—who could play higher, faster, and, most importantly, correctly? After three months of music lessons, teachers started to comment on Dillon's reading skills. They were improving steadily and some surprisingly large leaps had happened, which they put down to a new school, new start, and more support. But then one of the teachers mentioned that a boy at a very similar level arrived at the school at the same time and they weren't seeing the same positive changes in him.

Of course, it would be easy to put these positive changes in reading down to Dillon learning how to read music and that would be a nice ending to the chapter. But if I have learned anything from this research and its application in schools it's that every child's brain is incredibly individual and that developmental progress is made through changes in our neural wiring and our environment, and is influenced by our genetics and predispositions. Dillon had the modeling of a peer who was motivating through friendship and daily small goals to benchmark himself against. He may have had some interesting wiring, just like me, that was short-circuiting the reading process. He may also have had an auditory processing issue that exhibited itself through his speech. The beauty of learning music in childhood is that the experience itself, with all of its facets, can improve, amend, rewire, and refocus cognitive development in ways we could never have known were needed.

Music learning is not a single problem, single solution intervention or experience. A child's brain will take what it needs to grow from the multidimensional experience—all we have to do is give it the opportunity. Last time I saw Dillon he barreled up to show me the first chapter book he had ever finished reading, and he was so excited about all the new facts he knew about dragons.

Further reading

Hansen, D., Bernstorf, E. & Stuber, G.M. (2014) *The Music and Literacy Connection*, Lanham, Maryland: Rowman & Littlefield.

Moreno, S., Friesen, D. & Bialystok, E. (2011) "Effect of music training on promoting preliteracy skills: Preliminary causal evidence," *Music Perception: An Interdisciplinary Journal*, 29(2), pp. 165–72.

Patel, A.D. (2010) *Music, Language, and the Brain*, Oxford: Oxford University Press.

Schellenberg, E.G. (2004) "Music lessons enhance IQ," *Psychological Science*, 15(8), pp. 511–14.

11

JUST PAY ATTENTION

How music stretches each child's attention span

My mother, a career elementary school teacher, used to have a saying: the formula for the attention span in a child is their age plus two minutes. When I share this phrase with elementary and early childhood educators there are always lots of bobbing heads in the audience. When I share it with pre-service teachers in my opening lecture, they are horrified. When I share it with parents, they roll their eyes and chuckle.

My mother did her best work with children aged around six and seven years, particularly those who were struggling with their reading skills. She would say that the first and biggest hurdle was often the students' short attention spans: If they couldn't pay attention for long enough to read a full sentence, how could their reading progress? She would also talk about their problems with memory for letters and language sounds. As I delve into the research about music learning and how it relates to attention skills and working memory, I recalled her formula and the impact that the underdevelopment of these two skills of attention and memory had on children's learning in her everyday experience.

Attention is an interesting skill. It isn't an on or off skill—now I'm paying attention and now I'm not. It is constantly fluctuating depending on our environment, our own base attention skills, and how well we are functioning, for example, if we are hungry or tired. For children, attention is both a born and learned skill and develops with age. In a school setting I might hear teachers and parents say "He just struggles to pay attention" or "She could do much better but she gets distracted so easily." Then there's the very common report comment: "He could achieve a higher grade but his lack of attention prevents him from achieving his potential."

Ringing any bells? Were you a "could do better with improved attention" child, or is your child one? If you are a teacher, how much of your own attention in class is taken up by those students who struggle to pay attention? This chapter will look at attention in its many forms, how it develops and how it is impacted by many factors. The natural partner of attention is working memory, which is the capacity to temporarily hold on to information for processing. Both of these skills are vital for effective learning and both have been found to exist in higher than normal levels in musically trained children. Let's find out why.

Just pay attention

Attention has many definitions but the simple one that I like is the ability to choose what to focus on and what to ignore. This can be a hard decision sometimes because the thing you are meant to be paying attention to might not be very interesting, and the shiny thing over there is very interesting. This is even harder for children to do as their shifting attention, their inquisitive nature, is their prime avenue for understanding the world in the first seven or eight years. This natural curiosity is the basis of the play-based learning and inquiry learning that is now being used in schools all over the world as an alternative method for children to learn based on exploring and asking questions that lead to more questions. Such methods have an enormous capacity to instill a love of learning in every child.

This approach to learning can be contradicted by the "just pay attention" mantra that is often pervasive in mainstream schools after seven years of age. It is the industrial approach to education that British education expert Sir Ken Robinson so eloquently rails against when he says schools are killing creativity and a love of learning: "If you sit kids down, hour after hour, doing low-grade clerical work, don't be surprised if they start to fidget." There is so much understanding now about how rote learning, in which children amass knowledge for the purposes of a test and then promptly forget it, is probably not going to work for the next generation of adults. So we are at a crossroads with much of our educational theory, and while I could write a whole different book on that, let's look specifically at how music learning relates to attention.

Our auditory processing system, the processing of all of the sound around us for information, is our greatest information-gathering sense. What I mean by that is it collects the most information, it does so constantly, and it never switches off. I've mentioned that I regularly ask audiences which of their senses they believe gathers the most information. School leaders, nonarts teachers, and parents always say eyes. Music teachers and musicians always say ears. These answers intrigue me. I am not sure if the first group are working off their own experience as well as previous educational experiences that focus on seeing information first, but the latter have spent decades honing their ears and therefore are more sensitive to sound. Whatever the reasons, research into musically trained brains has helped us understand just how important and pervasive our auditory processing network is.

It makes sense from an evolutionary perspective. We need to know if there is a bear in our cave, and the only way we will know when we are asleep is through sound. If I use another example, have you ever woken up during the night for no apparent reason but then you replay the sound in your head that made you wake up, a sound that is out of the ordinary for your sound environment? That's called echoic memory, or "there is a bear in my cave" memory. Our ears never turn off from processing information and the amount of auditory information is increasing exponentially.

Our world has never been as noisy as it is now. The environmental noise our auditory processing network has to continually process is far beyond what it was built for. Consequently, the ability to pay attention or maintain our attentional control on what we want or need to focus on is getting harder. Essentially, we are often asking our children's brains to learn even more than we did in school while, often unwittingly, placing ever more auditory distractions around them. Unfortunately, at the same time, in many school settings children are getting not more but less access to the very activity that hones their auditory processing network: music learning. Our everyday environment is unlikely to get quieter, and with this new knowledge about how important auditory processing is to learning it is important to give our children the cognitive tools to handle the world they will inherit.

Attention is not an on/off switch

Attention is not an on/off switch but instead a very complex series of processes and decisions made moment by moment, and often unconsciously. Goal-directed processes are one of those series of skills. Goal-directed processes ask us to maintain our attention on a goal. This could be as simple as getting the round peg into the round hole for a toddler, to finishing a five-word sentence for a five-year-old, to finishing a university degree for an adult. These goals can be interrupted by a conflict, like a noise outside for the toddler, a student talking next to them for the five-year-old, or a life event like an illness for the university student. Our attention to these goals can be interrupted very easily, by an external distraction through to an internal frustration, and reigniting that attention after you have lost it can seem like a monumental task. Anyone with a job that includes answering phone calls while doing other tasks will have thought to themselves, "Now, what was I doing before that interruption?"

In spite of what the phrase "just pay attention" might imply, paying attention is a constantly changing requirement in our brains and there are many theories about how we shift, process, and reengage our attention. To give you a taste of these theories, take the concept of early and late attention selection theory, which suggests that the order in which we process information might both inform and shift our attention. When we pay attention to a car driving by, our early attention may take in the sensory information, the sound, smell, or movement of the car, while our late attention may take in certain features such as color, make, and model. Dr. Psyche Loui, an emerging researcher in the neuromusical field, described attention as acting on multiple levels, gathering and integrating sensory information and turning it into cognitive understanding while at the same time placing it in a context or associating it with other information. How and what we pay attention to is an incredibly complex, multidimensional, mostly subconscious, and highly individual cognitive process that I imagine to be like a brilliant sphere of activity in our brains, not dissimilar to the fireworks the researchers saw in the first fMRI scans of participants listening to music.

The ebb and flow of attention

One of the earliest discoveries in the field of neuromusical research was that professional musicians seemed to perform better on tasks that measured attention. Often these tasks were part of a series of tests into executive function of which attention skills were an important part. There has been a lot of debate about why playing and learning an instrument might result in heightened attention skills. Could professional musicians have a predisposition toward maintaining attention and are therefore able to practice their instruments for longer and consequently become better at them? Could it be that the act of learning and playing music itself improves attention skills? Or could it even be that music is a vehicle to enhanced attention and there is no deliberate learning or training required? As you have heard me say often by now, it is probably all of the above to some degree, and this is because every child is born with different dispositions and their development involves different influences and experiences. What we do know is that the nature of how music and attention are probably connected and possibly complementary is just beginning to be revealed.

Let's start with music itself. When we listen to music we notice it is there, most of the time, and we may consciously acknowledge if we like it or not, or if it changes our mood. But what we probably don't know is happening is that our brains are paying a lot of attention to the music and simultaneously separating and reassembling the sounds together. We don't need to learn music to do this, but those who do learn music are sort of like mindfulness gurus of music. They are both aware of how their brains are processing the music and they can often put labels to the different parts of the music, like the instruments, the speed, or the bass line. To parallel a Buddhist meditation concept, it is the ability to label a thought as thinking, to stand back in our minds and see a thought and watch it float by without engaging with it. Musically trained people can hear the music float by, label all of its parts as melody, harmony, rhythm, and so on but can choose to be emotionally affected by the music or not. To do this, as a Buddhist or a musician, requires heightened levels of attention and cognitive control.

Our brains process music both as a moment in time and a continual stream of information. Think of a song you heard today. Imagine in that brilliant sphere of activity that is your brain you were listening for the melody of the singer, which is happening over time like a horizontal line, while simultaneously relating it to how the drum and bass lines, like a vertical line, were complementing the melody. Then your brain was also listening for repetition or something a bit unexpected in the music, predicting and expecting what comes next. Our brains love it both when our predictions are satisfied, like a reward for knowing what comes next, and when something expected happens to tweak our interest. As Dr. Loui describes it, music listening is the analyzing of a busy auditory scene, and that scene requires nuanced and persistent attention.

Why can musicians pay attention better?

As I mentioned above, there may be more than one answer to this question. We know through the explosion of neuroscientific research over the last two decades that development is impacted by both what we are born with, our predispositions, and what we experience, our environmental factors. The division of that influence is still being hotly debated but I like to express it as not nature *or* nurture but nature *and* nurture. Some children are born with high levels of attentional control which, when applied to music learning, can make it look really easy because they aren't struggling to maintain their attention. One of the most profound occurrences I have seen of this as a teacher was when a fourteen-year-old boy who had never played an instrument before, or even been interested in one, got the opportunity to try the clarinet in a music class. By the age of eighteen he had worked through all the exam grades that usually take eight to ten years to complete. He fell in love with the instrument for sure, but I suspect he also had the attentional control to maintain that love through the tougher practice days.

However, music learning has also been found to improve attention skills, often from a very low baseline, which can be a particular challenge for students with ADHD or autism. I have had the great privilege of helping a number of schools establish music programs now. The schools might be implementing music programs to develop their students' language skills or maybe their inhibitory control, but without fail one of the first changes they see in the students is their ability to maintain attention for longer. It's an important step because to see improvements in academic learning, some students first need to learn how to pay attention for longer and to ride the attention roller coaster a little better. Visiting one of these schools, a fifth grade teacher made a beeline for me and took hold of my arm with a sense of great urgency, a huge smile on her face. "I just have to tell you," she said, "Marcus finished his math worksheet! First time ever—before, he couldn't even get to the second question—but he did it and he was so proud. I was so proud I could have cried. I wouldn't have believed it but the only thing that is different is the music program."

What would it have been about the music program that might have improved Marcus's ability to finish a math sheet? The first thing may have been an accelerated maturity of auditory processing, which means the ability to activate that brilliant sphere of activity in a more nuanced and deliberate way. Think of it as being able to taste all the five spices individually in a Chinese five-spice mix. This kind of attention requires cognitive control, and that is a core skill in music learning.

The second thing could have been the music program's requirement for precise motor control and coordination of the motor, visual, and auditory cortices. This not only requires attention but coordination or entrainment at a very exact level. I still love being present for the moment when a young child has been learning an instrument for a few months and suddenly that coordination all falls into place and they seem to zoom ahead.

The third thing that could have made a difference is experiencing a reward for maintaining attention. So often children give up on a task if they can't see or feel a sense of achievement or success, even when they are trying very hard. The gap between "try" and "give up" gets shorter and shorter with every experience of failure. Why would we keep trying when we are pretty sure, through previous experience, that we will fail? It doesn't take long for a child to stop trying altogether. But music learning seems to short-circuit that message. This may be because of the micro goals of just getting the right sound out of the instrument at the right time or because of the social nature of music learning, which makes us feel like we are all in it together. It might be different for every child but what music teachers know and researchers have also been discovering is that music learning enables each child to improve their attention skills in their own way.

Music five-spice

Professor Aniruddh Patel could be described as the father of music and language research and has an incredible gift for taking very complex ideas and expressing them in elegant and accessible ways. His OPERA hypothesis has five parts (hence my repeating five-spice references, for which I hope he'll forgive me) and is a groundbreaking theory about how music listening and learning uniquely enhances the human brain, specifically in the areas of language and music. OPERA stands for Overlap, Precision, Emotion, Repetition, and Attention. Professor Patel suggests that music and language use an *overlapping* neural network that processes sound characteristics of both music and speech, that music training requires *precision* in processing information and producing music, that music elicits positive *emotions* and engagement, and that music learning requires *repetition* and sustained *attention*. It is a mix of cognitive activities that are both hard to combine correctly but so satisfying when we do, both as a musician and as a music listener. This high-stakes, high-reward combination may be the essence of how music learning enhances brain development in such a unique and ultimately beneficial way.

Sprinkling the five-spice around

If you've got a good metaphor, stick with it. To get the right combination of spices it is a case of trial and error, repetition, and attention to the changes that occur when we try a new combination. As both parents and teachers we sometimes forget that learning anything is not a steady gradation of skill development toward mastery but a period of satisfying improvement followed by a plateau of frustration. To engage an overlapping neural network in a precise, emotionally controlled, frustratingly repetitive, and attention-sucking way is hard work, and some days are not diamonds. But to get to that satisfying improvement we need to help children understand the plateaus and how paying attention to these as much as the successes is an equally important part of the process. And doesn't that sound an awful lot like the development of resilience?

Further reading

Besson, M., Chobert, J. & Marie, C. (2011) "Transfer of training between music and speech: Common processing, attention, and memory," *Frontiers in Psychology*, 2, p. 94.

Habibi, A., Damasio, A., Ilari, B., Elliott Sachs, M. & Damasio, H. (2018) "Music training and child development: A review of recent findings from a longitudinal study," *Annals of the New York Academy of Sciences*, 1423(1), pp. 73–81.

Kraus, N. & Chandrasekaran, B. (2010) "Music training for the development of auditory skills," *Nature Reviews Neuroscience*, 11(8), p. 599.

Loui, P. & Guetta, R. (2018) "Music and attention, executive function, and creativity," in M. H. Thaut and D. A. Hodges (eds), *The Oxford Handbook of Music and the Brain*, Oxford: Oxford Handbooks.

Patel, A.D. (2011) "Why would musical training benefit the neural encoding of speech? The OPERA hypothesis," *Frontiers in Psychology*, 2, p. 142.

Robinson, K. (2006) *Ken Robinson: Do schools kill creativity?* <www.ted.com/talks/sir_ken_robinson_do_schools_kill_creativity/transcript?language=en>, accessed July 18, 2019.

12

COMFORTABLE WITH DISCOMFORT

How music teaches your child to love frustration

"Nailed it!" says Amber, a young percussion student, to her teacher with a beaming smile. She has just finished a two-bar rhythmic pattern on the drum that she has been struggling with for five minutes, but it seems like forever to this seven-year-old.

"Great, well done! Now do it again," says her teacher with a knowing grin. From years of music teaching experience he knows that the ability to repeat a pattern correctly is where the real learning happens. He also knows that his excited student needs to get to the point where the pattern is almost automatic for the cognitive click to really have settled into place.

The "nailed it" experience in the brain

Let's go back to the moment when the student "nailed" the rhythm. If you aren't familiar with the term "nailed it," replace it with "I got it." Think of that moment you shoot the scrunched-up piece of paper into the bin or, if you are like me, the moment you arrive at a meeting right on time having coordinated planes, trains, and automobiles to get there. I find the use of the phrase "nailed it" interesting because, if we take it literally, it means fixed in place, and maybe this is not too far from the actual cognitive experience that is going on in that student's brain. Let me explain further.

In the moment when the student "nailed" the rhythm, a rush of dopamine went through her brain and body. The reward network is helping the brain connect between the activity and the feeling of pleasure, which then makes us want to repeat that behavior. We know this through studying addiction in the brain, which is the overactivation of the reward network, but also through studying how the human brain processes music and music learning.

The initial "nailed it" is more than just a good feeling—in music learning it is a combination of coordination and control of her arms and body to move at exactly the right time to make the correct sound that matches the prediction. It is the cognitive synchronization of the eyes, ears, and body as she reads the music, hears it in her brain, and re-creates it through her body, and then checks back with her ears. It is the mastery of her emotional state; she has been at it for five whole minutes and during this time has felt her temperature rise through heightened frustration and has had the fleeting urge to get up and leave the room. While she has wanted to nail it for her own satisfaction, she also wants to please her teacher and show him that she can complete the rhythm that the teacher seems so sure she can do. My students often tell me I am asking them to do so much at the same time, and when you break it down like this, I guess maybe I am.

In many ways, this is the beauty of music learning on an instrument. It is all of this coordination, synchronization, personal and social motivation, and ultimately moments of intense reward all wrapped up in a bow. This is a perfect storm for the release of dopamine in the student's brain to help her cope with the anxiety and stress this task will be causing her. The very nature of music learning means all of these things do need to take place at once, but what happens in the brain is that through the challenge of music learning it rises to the occasion by "nailing it" in terms of higher cognitive function.

Our own personal drug lab

You might not like this idea, and certainly when I speak to principals they squirm a little, but in the end, our brains are one of the most effective drug producers and couriers around. The trick is to get the good stuff, the good drugs that aid our development, not hinder it.

One of my favorite sayings is "we learn what we care about." The sentiment has been attributed to everyone from Gandhi to Goethe, and in the neuromusical world it is a touchstone idea from which the study of reward networks has grown. Let's go back to our nailed-it moment: a rush of dopamine, a little bit of disbelief that after trying for so long it all seemed to just click into place, but most importantly, a trigger for connecting a specific activity with a feeling of pleasure, and a pleasure she wants to re-create and seek again. Put another way, it is the natural production of a drug that makes us feel good and we want that internal drug lab to make more. If I switch it to educational speak, this is the same concern, in a different context, of the lifelong learner that teachers are all trying to help create, the learner who goes out into the world hungry for learning.

Back again to our nailed-it moment. While I started with the time of endorphin release, the satisfaction level of that experience was made higher by the five minutes of frustration and failure that came before the moment of triumph. A success is made even sweeter by the level of trial, error, failure, and difficulty that had to be overcome in the realization of that success. Built into the process of learning to play a musical instrument is a small but measurable level of trial, error, failure, and difficulty. Wrapped up in the music learning bow is the experience of getting it wrong more than getting it right and microdosing on frustration every time our percussionist picks up her drumsticks.

But she is also microdosing on success and the dopamine that comes with that. Amber has achieved physiological, cognitive, and social success in a tiny moment of triumph. Now for the real nailed-it moment: she has to try to nail it again.

Can you guess what usually happens? If you said she tries again and doesn't nail it, you're right. She scrunches up her face, maybe even looks a bit crestfallen because she can't replicate the rhythm she just perfected. But then a new fountain of dopamine kicks in and she is determined to get it on the next try. She focuses her cognitive energy, all that coordination, synchronization, and motivation, to achieve this one goal. And in most cases, third time is a charm.

So what does this process teach her? The five minutes of frustration teaches her that by just sticking with it, eventually you will get it. As an adult I know some of you might be thinking that no matter how hard you try there are some things you just won't get. But children in their prime learning phase of life don't know that and they need every opportunity, and in some cases to suspend their disbelief, in order to achieve what they are capable of. Children don't know their limits yet, and often the most rewarding part of teaching is helping them achieve something they didn't know they could do.

The lack of success on the second attempt to nail it teaches Amber that doing it once does not totally cement the process for her. She needs to repeat it and get it wrong more times in order to get it right more regularly. A wonderful educator in the neuromusical field named Liisa Henriksson-Macaulay summarizes this part of the learning process well: "Music training for the young brain is like building motorways where dirt tracks would have been." The first time Amber nailed the rhythm she made the first impression of the dirt track. One touch of rain or left unattended and the track would be gone in a day or two. To create a highway from that dirt track, Amber would need to repeat the rhythm many times and also play that rhythm as part of a longer piece. Then the highway would be both created and maintained and become a permanent highway in her brain. If I think about the number of rhythms Amber will learn and combine, that is a pretty amazing highway system racing around the cerebellum.

Microdosing on music

This phrase could be expanded to microdosing on frustration through music learning. Apart from the wrapped-in-a-bow nature of music learning, the brain's reward network has a very particular reaction to sound processing. Our largest information-gathering sense is our hearing, and by extension our auditory processing network may be one of the hardest working networks in our brain. Our auditory processing takes in the most sound and processes it for details, which could be anything from a sound that we need to pay attention to for our own safety, a change in sound that indicates a change in what is happening around us, or a filter for language and how anything from the syntax, prosody, or tone of what someone is saying may have changed. When Professor Nina Kraus from Northwestern University in Evanston, Illinois, explained this to me, I walked out of her lab that chilly day hearing and viewing the world entirely differently. I marveled at how much my brain was processing all the time, mostly without my conscious knowledge. I now am very thankful for my auditory processing when I travel to new places. It is keeping me safe and informed every minute of my travels.

When the reward network is activated through music learning it has an influence on the auditory processing network, which in turn influences many other functions and structures in the brain. A theoretical framework put forward by Professor Kraus is that the reward component of music learning may promote the reorganization of the structures within the auditory processing network, inform auditory learning, and create conditions for learning beyond auditory processing, which might be the transfer of skills beyond that auditory network.

The reward network is very complex, releasing dopamine through multiple channels including the mesolimbic and mesocortical pathways. The reward network has been found to be heavily influenced by auditory processing, particularly in a social context. This has been discovered by studying music learning in music ensembles, because it is a double hit of reward. We know that around ninety-five percent of people fire their dopamine pathways when they listen to and learn music, and this is magnified when they are in a social context like a music ensemble such as a choir or rock band. When the reward network is activated we feel pleasure but it also sends a message to our brains that whatever is happening is something that is worth remembering and repeating.

Professor Kraus and one of her co-authors, Travis White-Schwoch, go further and suggest that the study of music learning may have contributed to the discovery of the brain mechanism that creates the "we learn what we care about" experience. They call it the cognitive-sensorimotor-reward framework. I would translate that to the thinking-feeling-feel-good framework. It is an elegant and complex framework that essentially captures the idea that when children experience learning through their sensorimotor network it creates both a reward network reaction and a cognitive network reaction that enhances the brain's learning capacity. It is like the perfect cocktail of internally made drugs for learning, wrapped up in a music learning pill.

Do you want the red pill or the blue pill?

I knew I had to invoke *The Matrix* at some point in this book. This 1999 movie captivated action buffs with its incredible special effects and fight scenes and the minds of philosophers and existentialists with its profound questions about the human condition. In short, the protagonist, Neo, is offered an opportunity by the rebel leader, Morpheus, to either remain in the familiar but constructed version of his world by taking the blue pill or be given the opportunity to see the world in a different and possibly more real way by taking the red pill, a journey that would come with challenges and confrontations.

Stay with me. I am not saying children should have red or blue pills or that music learning is a choice between a make-believe life or a real, far scarier world. However, I do think we need to look again at what we believe about music learning as an experience and the messages we give our children and students about the music learning process.

"Sam isn't having fun learning trumpet anymore, so I am going to let him give it up." This is an all too familiar conversation that I and many other music teachers have with parents every year. It is one I struggle with because no teacher would ever have said that music learning was predominantly fun—honestly, it is predominantly hard work. As a day-to-day experience in getting things wrong more than you get them right, it is an ongoing small frustration because there is no end to it as there is always the next piece to learn or the next note to get right.

However, this is the sort of hard work that the brain thrives on. It leads to the reward network firing up the cognitive and sensorimotor network, which in turn ups the drug production of endorphins and dopamine in almost every child's brain. Musically trained children love learning—they crave it even. There have been a number of studies that have searched for the right musical dose for permanent brain enhancement, or, put another way, how much music learning do you need to permanently change the brain of a child and possibly create this lifelong love or craving for learning? You can imagine this is a very difficult thing to measure but the thinking at the moment is it is more than two years of continuous learning before the age of seven, and three years of continuous learning after the age of seven (remembering the sensitivity period for learning that has been identified from the age of birth to seven years). One excellent study led by Dr. Diana Zhang found that across ninety-five studies a "six-year rule" emerged as the period of music learning time that classified a musician, but it does not necessarily follow that this is the time required for permanent brain enhancement. Whatever the magic number might be, and that is still very much up for debate and exploration, as a teacher I have observed that musically trained students love the next challenge and not only embrace struggle and challenge but continue to search it out in life well after their schooling ends. This attraction to being a lifelong learner could be connected to the amount of time and the type of music learning experiences children are given in their years at school.

What may be even more important as we progress through this century is the fact that musically trained children have sat with and learned to understand and use frustration as a tool for learning. It is only human to shy away from what is hard and causes frustration. While we talk about how the messy middle is where we do our best learning, when we are smack bang in it no one likes how it feels. The difference is having experience with the feeling of frustration, because being comfortable with discomfort makes it less daunting and difficult to manage. If you think about what life is like for Amber at school each day, she might have six lessons on different areas of learning, all of which have varying levels of frustration and the need for persistence built into them. That is what learning is, the act of continually encountering new ideas, challenges, or activities that, by their very nature, are uncomfortable and frustrating in the beginning. The ability to stick with it while managing these very distracting emotions is the mark of an effective and lifelong learner.

I respect parents who obviously consider the pros and cons of allowing their child to give up learning an instrument. The statement on page 143 does not take into account Sam's age, how long he has been learning, if he is heavily committed to other activities outside of his music learning, the fact that he might be outgrowing his instrument or that he might not feel challenged or engaged in learning by his trumpet teacher. But taken as a single point in time, I struggle with the idea that when music learning stops being fun it is time to stop learning. This to me is the blue pill, the pill that allows the learning to end along with the potential to build on the networks that will help Sam conquer many more challenges than just music in his later life. Sometimes, and I choose my parents carefully, I may come back with a question for them: "Okay, if you let Sam give up learning trumpet when it becomes difficult, what is he actually learning?"

The red pill, the Wonderland as Morpheus calls it, is the real but more confronting world. It is the choice for Sam to continue to learn his trumpet but to address the level of discomfort and frustration he has with the learning process. While Sam probably said to his mom or dad, "I hate trumpet and I don't want to learn anymore," what he might have hated far more was the lived experience of consistent frustration. This is a deeper world, a more real world in some ways, because it speaks to Sam's persistence and resilience as a learner. One way to address this is to give words to the feeling. Most of the time with my students I see one reaction that is hiding a deeper fear, that when we are young might be hard to pinpoint or name. Asking questions like "Where does it live in your body?" or "How does your body (temperature, muscles, sight) change?" begins to help Sam identify when it happens again. Another way is to bargain one hard thing for one easy thing, which could be "Do one hard thing your teacher assigned to you for five minutes and then do one easy thing for ten minutes." This helps to activate the reward network both for doing hard and challenging things and having the freedom and choice to do what you like.

Music learning is a present, wrapped up in a big red bow. By its very nature music learning is being understood by researchers and educators as an experience that promotes brain development through activating and challenging the brain to grow in a multidimensional way. In turn, that multidimensional growth promotes the rounded, expansive, and nuanced growth of a little human who, given the opportunity, will learn to love the feeling of frustration.

Further reading

Blood, A.J. & Zatorre, R.J. (2001) "Intensely pleasurable responses to music correlate with activity in brain regions implicated in reward and emotion," *Proceedings of the National Academy of Sciences*, 98(20), pp. 11818–23.

Henriksson-Macaulay, L. (2014) *The Music Miracle: The scientific secret to unlocking your child's full potential*, London: Earnest House Publishing.

Hyde, K., Lerch, J., Norton, A., Forgeard, M., Winner, E., Evans, A. & Schlaug, G. (2009) "The effects of musical training on structural brain development," *Annals of the New York Academy of Sciences*, 1169(1), pp. 182–86.

Kraus, N. & White-Schwoch, T. (2015) "Unraveling the biology of auditory learning: A cognitive–sensorimotor–reward framework," *Trends in Cognitive Sciences*, 19(11), pp. 642–54.

Zhang, J.D., Susino, M., McPherson, G.E. & Schubert, E. (2018) "The definition of a musician in music psychology: A literature review and the six-year rule," *Psychology of Music*, 0305735618804038.

13

SPRING-CLEANING YOUR BRAIN
How music helps brain wiring

My daughter was born about a third of the way through my PhD study. Here I was, researching brain development in children, and now I had my own experiment participant! Just to be clear, I never experimented on her in any way, but through simply observing her development and that of her peers I had a perfect opportunity to connect the research to real life, and to let real life teach me more about the research.

One concept that I learned about shocked and even saddened me a little. It turned out that my daughter's brain would experience something called synaptic pruning, a process where our brain does some serious spring-cleaning of all the information pathways in our brain. They do this to make space for new information and to make the existing pathways more efficient and easier for information to travel along. But the name just made me think of a pair of sharp shears taking to the connections within my daughter's brain like an unruly rosebush.

The pruning process is perfectly normal and something that we cannot control at all. It is actually one of the many helpful processes of neuroplasticity that maintain the health of our brain. Any cupboard needs a good cleanout every so often, and afterward that cupboard usually functions a lot better. As far as our brains go, this pruning is occurring all the time to a small degree. It is a process that is active all the time like an assistant following around after you, picking up memories, organizing thoughts neatly and storing useful new facts in a place where we can easily find them and use them again (or discarding them, as the case may be).

Research conducted in the early and mid 2010s pointed to a sensitivity period for learning from 0–7 years. As we enter the next decade this concept of a sensitivity period is being challenged and refined. In this chapter I will refer to the first seven years of learning but I remain aware that, as always, research is forging ahead in this area.

Children are like a sponge

The phrase "children are like a sponge" took on a whole new level of meaning for me as I learned about synaptic pruning. The concept of sensitivity periods in cognitive development, which means periods in our lives where we are more open and capable of learning new things, fits with the sponge saying. In the first seven years of a child's life they seem to absorb everything around them, mimicking speech and behavior through close observation. They learn how to speak a language or sometimes two, a herculean cognitive task that appears to happen so easily and naturally from the outside. While pruning is still occurring, their brain sponge is taking in more water than it is squeezing out.

At some point our brains reach capacity, having soaked up so much new information that they reach the full line. This is when researchers believe a significant period of synaptic pruning occurs, the difference between the general house tidying we might do on a day-to-day basis versus the weekend you hire a huge trailer and do a serious cleanout of the garage.

Two things tend to happen when we do a major spring-clean. The first is that the whole world really needs to stop: you have one task to do, the clock is ticking on the trailer hire, and you need to get the job done. Children's brains are the same and they kind of slow down a bit in their learning in order to get this big job done, and this is mirrored in the pla-teauing of learning or dip in test scores that we see in children around this age. The second thing that tends to happen is a bit of cleaning detachment or indifference—sometimes we just start throwing everything out without thinking of alternate uses for it or how much we value it. In a child's brain going through significant synaptic pruning, this might be memories, thinking pathways or previously useful message routes. This could also slow their learning down in unpredictable or unusual ways for a period of six to twelve months.

Music learning is an excellent way to study this phenom-enon of human development because it often starts early in a child's life during this sensitivity period of learning. The circuitry in the human brain is highly plastic between birth and seven years of age, and in many cases the neural pathways that will have been there from birth will be getting their first workout, which provides an opportunity for them to expand and become more stable and reliable. The reason music learning is a useful tool in better understanding synaptic pruning and neuroplasticity is because it demands a lot of work by the brain. In particular, learning a musical instrument involves highly complex tasks across multiple modes of the brain. It requires high levels of connectivity and development and therefore is thought to both enhance and promote neuroplasticity. This is just as true when the brain is growing connections and learning as when pruning connections and consolidating. It is a tool for the brain to continue to function effectively, just like having the best assistant in the world to help you run your life.

Wiring right the first time around

We are all born with complex neural wiring already in place. This wiring is influenced by genetic, personality, and environmental factors, and possibly other factors during pregnancy. After birth those wirings get used, tested, changed, connected, strengthened, broadened, rerouted, abandoned—there are just so many possibilities. It has been suggested that the first seven years of life is probably the initial opportunity for this wiring to occur. Wiring is probably a very crude word for the incredible way the brain forms connections and develops, but it is a starting point in understanding the brain and how it grows.

Learning music promotes connection between multiple parts of the human brain. That connectivity is between the brain buildings, the structures within the brain, and the brain

roads, the functions within the brain that connect the buildings together. Like any city, roads and buildings change.
Sometimes this happens deliberately, like when we significantly remodel a building in order to change its purpose or
reroute and widen a road to cater for the amount of traffic
it now needs to carry. Music learning does this by engaging
multiple parts of the brain at once and then putting that connectivity through its paces. This may happen in a similar way
during growth but also during the pruning process. The daily
experience of learning to play a musical instrument, which
could be anything from a tiny eighth-size violin to keeping a
beat on a small drum, is similar to a quality assurance process
in a factory when you are trying to perfect your chocolate
sponge cake recipe: you change a little bit then test it, change
another little bit then test it. A product that is adjusted grad-
ually and tested along the way will always be better, stronger,
and more fit for purpose than one that isn't.

During periods of synaptic pruning having a touchstone
of picking up an instrument and testing out your newly wired
brain is very important for children. They may well know
that their brains are behaving a little differently, they may
know that a word they learned yesterday seems to have disap-
peared entirely from their brains today, they may notice that
their ability to control their impulses isn't quite as good as it
used to be. But a daily discipline that engages their multilevel
synaptic system can help them through this time. This idea
is not dissimilar to the Daily Mile in the United Kingdom,
where schoolchildren run at their own pace for fifteen minutes
every day, or the "brain run" that some Australian schools do
around the school perimeter. This practice warms up kids'

bodies and gets the blood flowing, with research showing that it improves learning. Imagine if before or after their run those same children warmed up their brain connectivity and primed their auditory processing network with ten minutes of music learning.

Losing in order to gain far more

I think the concept of synaptic pruning initially shocked me because in today's thinking, losing or letting go of something is inherently bad—we always have to have or be more, so why wouldn't it be the same with brain connections. It makes me think of movie plots built around the possibility of being able to access and use more of our brains, from which would spring the ability to learn to speak a new language in ten seconds. Wasn't this what we wanted, to be able to use more of our brains to cure cancer, solve cold fusion, and learn how to fly?

The answer is no. More is not better, and efficiency is what we are seeking when it comes to brain processing and connectivity, the shortest, fastest, and most flexible information pathway from A to B. I first came across this when reading research studies that compared groups of musicians and nonmusicians performing the same cognitive task, fully expecting to find that musicians used more of their brains to complete the task.

How wrong I was. Musicians in a large number of studies have been found to use less cognitive energy—which means less of the brain is being activated—to complete the same task when compared with nonmusicians. But not only did they use less cognitive energy to complete the task, they often completed it faster and with greater accuracy. It's like the saying goes: "If you want something done, give it to a busy person." Could it be that if you want a cognitive task done, give it to a busy, efficient, highly connected brain?

It is important to note here that the neuromusical research field has had some teething issues regarding how a musician and a nonmusician are defined. I would defend this by saying scientific research is the exploration of a phenomenon, and exploration has twists and turns. Early research using the musician/nonmusician model of research tended to have a loose definition of what a musician is, perhaps with very small participant groups of five or so who were professional violinists or pianists. They might have each had twenty years of music learning behind them, playing four hours per day from the age of five. These musicians were more than likely outliers in the musician world, talented enough to make a good living out of playing music, which is a very tough gig.

When these first research studies started finding that there was something worth studying, the participant groups got bigger and more varied. However, getting hold of participants can be hard, so these groups were often made up of college students, already outliers in a way as they had made it through school with good enough results to enter college. These bigger groups of participants also self-identified as musicians, which often meant their music learning experiences were varied, anything from the number of years they had learned (say, between one and ten) to how they had learned, which could be formally in lessons and band to informally in a church group. The variables were huge when it came to the scientific method.

Despite this, these studies still found evidence that this field of research was worth looking into more deeply, so researchers began to work backward and focus on children and the impact of learning music. Anyone who has conducted research knows that there are more layers of ethical considerations when it comes to studying children. Add this level of care and scrutiny to larger participant groups (in later studies, regularly more than a hundred musicians), more complex testing, and longer observation periods (from ten weeks to more than two years in some cases) and the scientific findings become more reliable. These were studies with children and their design still identified the musicians as children who had chosen to study music (called self-selecting) and were not randomly assigned to learning music. This is called a quasi-experimental approach, which means there is the possibility that musically trained children have better-connected and more efficient brain functions to begin with.

We have just entered a period of the gold standard of research study design in neuromusical research: the randomized, longitudinal study. These studies follow children's development for three to five years and are rigorous regarding the type of music learning that occurs. Children are randomly assigned to a condition, which means they might either learn music, a sport, or do no activity, and are tested prior to the start of the study for cognitive equity. So as much as we can make it with human beings, the children's brain development is comparatively equal. This means that from the baseline measure the cognitive development levels are as similar as possible, which is a process that can be achieved by excluding outliers (participants for whom cognitive measures are significantly different from the majority of students) or by doing rigorous statistical analysis on the baseline measures to make sure even the smallest differences are taken into account. From these sorts of studies we can now begin to determine what type and length of music learning impacts on brain development.

A recipe for brain development

A caveat: take the recipe that comes next with a very large pinch of salt. Research is an evolving process of discovery and enlightenment and what follows are just the headlines of what we have found out so far. It is all still subject to evaluation, argument, and interpretation. So, taken with the other research described so far, I will let you be the judge.

Music learning, particularly learning a musical instrument, has been found to change the human brain's development. This change is significant before the age of seven because of the sensitivity period for brain development that occurs during this time. Music learning consists of multiple types of cognitive information that is then processed at multiple levels and by multiple parts of the brain simultaneously. All this gathering and processing is done through the sensorimotor and cognitive networks and activates the reward network concurrently to induce a sense of challenge, achievement, and incentive to learn more, both when the brain is growing and pruning. It keeps the brain healthy, active, and hungry to grow.

All of this development does not come from just any type of music experience or learning. Stopping to listen to a busker in a train station is not music learning. Singing hymns in church every Sunday or bouncing along to the Wiggles in the car every day is not music learning. These activities may enhance auditory processing, set off the reward network, or tweak our interest in a new instrument sound, but none of them are music learning.

Music learning is a discipline, something we do every day or several times a week, that is challenging, hard work, social in nature, and has very high ups and many frustrating downs. While this is what is going on inside a child's brain, often music learning looks on the outside to be a whole lot of fun and just a little chaotic. Ironically, the more expertly led the music learning is at this age, the more seamlessly it looks easy and play-like.

The recipe for music learning that seems to result in permanent, positive brain development at this stage has the following ingredients:

- It involves singing, moving, and playing an instrument outside our bodies—this includes working to sing in tune, moving both spontaneously and to a beat, and playing an instrument that is age-appropriate (at two years old, that's a whole lot of pots and pans).
- It includes learning in both an individual environment and a group environment from an expert—this includes having a personal lesson with a teacher (which can happen in a group of three to five students, depending on the instrument) and in a larger group of maybe ten to as many as a hundred.
- It includes learning to read musical notation that is age-appropriate—music notation is just another language, a sound-to-symbol connection, but this connection can be made with a symbol as simple as a single line that means hit the drum once.

- It involves starting to learn before the age of seven years—this doesn't mean that children cannot start learning after the age of seven or that all is lost for your nine-year-old child and they will never get any benefit. Indeed, your brain right now could gain benefit from learning an instrument no matter what age you are. But if your young child can experience musical games, sound-to-symbol activities, and moving to music just once a week, you are on to a winner.

- It includes performance (but not necessarily in competition)—but this is for reasons other than the music. Music performance is a high-stakes experience, one that involves standing up in front of parents and peers and singing or playing music that your child has been practicing for weeks or months. It is about the nerves beforehand, the act of overcoming the fight-or-flight response and pushing ourselves to go out on stage, dealing with the unexpected things that happen during performance, the elation from making it through to the end of a song, and even the loss of all energy and ability to speak as the adrenaline stops flowing. Going through this process often and in a group of our peers helps children experience trepidation, nervousness, euphoria, and relief in an environment where they also experience success and support. Performing music trains children for performing in life.

- It is formal, sequential, and expertly led—too often we think that if someone can play a musical instrument they can therefore teach it. Any music teacher will tell you that the best performers often make mediocre teachers, mostly because they may never have needed to break down the process of music playing as explicitly as someone who wasn't so genetically predisposed to music in the beginning. Good learners make good teachers because they understand the learning process. Music learning that results in permanent, positive cognitive development follows a developmental path that music teachers have been figuring out through trial and error for centuries. For example, young children don't learn how to divide a beat until they have learned how to keep a beat, and a young clarinet player learns the notes with their left hand first before crossing over to using their right. Music learning is not magical, it is highly structured and sequential, and this knowledge comes from a skilled and trained teacher.

My daughter has made it through the major pruning period, I think, and I am delighting in the new, more nuanced, insightful and quirky connections she is making around knowledge, understanding, and learning.

Further reading

Cho, E. (2019) "Sensitive periods for music training from a cognitive neuroscience perspective: A review of the literature with implications for teaching practice," *International Journal of Music in Early Childhood*, 14(1), pp. 17–33.

Degé, F., Kubicek, C. & Schwarzer, G. (2011) "Music lessons and intelligence: A relation mediated by executive functions," *Music Perception: An Interdisciplinary Journal*, 29(2), pp. 195–201.

Habibi, A., Damasio, A., Ilari, B., Elliott Sachs, M. & Damasio, H. (2018) "Music training and child development: A review of recent findings from a longitudinal study," *Annals of the New York Academy of Sciences*, 1423(1), pp. 73–81.

Hannon, E.E. & Trainor, L.J. (2007) "Music acquisition: Effects of enculturation and formal training on development," *Trends in Cognitive Sciences*, 11(11), pp. 466–72.

Hudziak, J.J., Albaugh, M.D., Ducharme, S., Karama, S., Spottswood, M., Crehan, E., Evans, A.C., Botteron, K.N. & Brain Development Cooperative Group (2014) "Cortical thickness maturation and duration of music training:

Health-promoting activities shape brain development," *Journal of the American Academy of Child & Adolescent Psychiatry*, 53(11), pp. 1153–61.

Penhune, V.B. (2011) "Sensitive periods in human development: Evidence from musical training," *Cortex*, 47(9), pp. 1126–37.

Schlaug, G., Forgeard, M., Zhu, L., Norton, A., Norton, A. & Winner, E. (2009) "Training-induced neuroplasticity in young children," *Annals of the New York Academy of Sciences*, 1169, p. 205.

Tierney, A.T., Krizman, J. & Kraus, N. (2015) "Music training alters the course of adolescent auditory development," *Proceedings of the National Academy of Sciences*, 112(32), pp. 10062–67.

White, E.J., Hutka, S.A., Williams, L.J. & Moreno, S. (2013) "Learning, neural plasticity and sensitive periods: Implications for language acquisition, music training and transfer across the lifespan," *Frontiers in Systems Neuroscience*, 7, p. 90.

PART THREE

BRIDGING THE GAP BETWEEN CHILDHOOD AND ADOLESCENCE

9 TO 12–14 YEARS

14

CONFIDENCE IS KING
How music feeds self-confidence

In 2017 I was given the incredible opportunity to be involved in a television show called *Don't Stop the Music*. It was a three-part documentary that followed students, their parents, and a school as they implemented an instrumental and school-wide music program. It had all the ingredients of a great story: adorable and quirky ten- and eleven-year-olds, parents who never had the opportunity to participate in a music program themselves seeing their children transformed, a school principal who was working magic in a community of students and parents who faced many challenges, and a couple of international music legends thrown in to inspire the students. I pinched myself at the start of almost every filming day because I knew I was in the middle of something that would change the lives of those students and their community. What I didn't realize until after the television program aired was that we had given the Australian viewing public something that they would have had no other way of experiencing: an inside view to showing the way music learning transforms children.

When it comes to learning a musical instrument, parents and teachers generally only see the shiny, almost perfect music performance at an assembly or concert. This performance will most likely have come after anywhere between ten and a hundred hours of practice in the rehearsal room, which may have been supported by another ten to a hundred hours of practice at home. It often appears seamless, even easy, like when we see a pole vaulter at the Olympics sail effortlessly over the crossbar. But we probably have an idea about how many hours of practice that athlete had to put in to reach such a high level of performance. What is involved in reaching the same level of performance for a musician?

Don't Stop the Music allowed viewers to see through a window into the messy, exciting, frustrating, and exhilarating process of music learning. It isn't pretty most of the time. It sounds pretty horrible most of the time. It definitely doesn't resemble music as we know it in the first stages of learning a musical instrument. To the untrained ear, and this showed on the faces of many parents who attended the brass band rehearsals from the beginning, it seems like a noise that could never turn into music that is enjoyable to listen to.

But the thing is that it always does, every time. Time, patience, expert instruction, and a hell of a lot of positive reinforcement always leads to a successful, shiny music performance in the end. What I realized when I had to articulate this process on the television show is that I have the great privilege of seeing this happen multiple times a year in schools all around the country and around the world. It didn't seem wondrous to me, just expected, but it was a true wonder to the parents and teachers involved. That part was magical, and it all starts with confidence. In this chapter I hope to help you see through that window, too, the window of how music in all of its ugliness at the start becomes an incredibly wonderful experience in the end.

Something has changed

I like working in schools that educate children from challenging backgrounds. I always struggle with the "right" terms to use—disadvantaged, underserved, challenging, or even traumatic backgrounds. What I mean are schools where external factors from family and socioeconomic situations create greater challenges for kids when they come to school. The reasons I love working with these schools are threefold: the school leaders are usually incredibly innovative and committed people, all the staff from teachers to cafeteria workers are open to whatever works for the children, and the students themselves benefit more from a school music program than their middle-income and well-off peers.

I get to work with these schools over a period of anything from nine months to three years. *Don't Stop the Music* was a supercharged version of the implementation of a music program. This was because a filming timetable provided the impetus to work out a lot of the implementation issues quickly, there was both the goodwill and the financial support to get it up and running effectively, and there was advice and constant support from educational experts. But in the schools I have worked in that didn't have this level of help, the same thing happens about three months after the instruments arrive and the music learning commences in earnest. The principal and often other staff members come to me and say something has changed.

What they describe can be anything from an otherwise disengaged and disruptive student, let's call him Isaac, finishing a whole math worksheet in a single lesson, to a nonverbal student, let's call her Mika, who had not spoken since witnessing an incredibly traumatic event, beginning to talk in her music lesson. These stories are simultaneously heart-wrenching and heartwarming, but they all have a common theme underpinning them that can often be attributed to the music program: they find their confidence.

Confidence is a tricky concept. We often know it when we see it, but try to explain it in words and you might struggle. When I ask Dr. Google to define confidence I get "the feeling or belief that one can have faith in or rely on someone or something," and this gets me thinking about these students in particular and the fact that relying on someone or something is actually not a given for them. In many of these communities the students have never known an adult who has maintained regular employment, some of them don't know where or if they will have dinner each night, and many of them deal with the consequences of drug abuse and domestic violence on a daily basis. How does music learning provide a belief that they can rely on themselves or others?

The gift of music learning

I use the word "gift" deliberately because, as I mentioned earlier, music learning is an experience that, wrapped up in a bow, provides so many different positive effects. This is not an area of exploration you can easily pinpoint in the neuroscience, psychology, or educational research because confidence is a very hard thing to measure as a concept, mainly because it can be the result of many experiences. Teachers are often the best-placed people in a child's life to identify when confidence has changed.

Isaac is a great example. This particular student struggled with his attention span and often his disruptive behavior was his way of hiding that he didn't understand the work he was being asked to do. Then Isaac joined the music learning program at his school on the violin, three small group sessions a week at school with a teacher and an ensemble rehearsal with fifty of his peers playing open string (what a glorious sound that makes). Six months after starting he completed his math worksheet for the first time. I remember this story vividly because it was his class teacher who ran up to me in the playground, put her hands on my shoulders, and joyfully but gently shook me with the news of his triumph.

What had happened for Isaac in those six months of learning violin? Most probably the consistent discipline of paying attention in music class extended his attention span generally. In Chapter 11 we explored attention in more detail and an example given was about Marcus who was struggling to know what paying attention felt like. Coupled with this may have been his new experience of try and try again, getting the note wrong more often than getting it right, but being urged to keep trying and doing this with his peers. Possibly Isaac may also have had some underdeveloped or underperforming brain connectivity that was improved by learning violin. Of course, we won't know for sure without rigorous testing and he is only one student, but for his teacher, the ability to complete a math worksheet is a success that will lead to other successes. The possible combination of attention development, inhibitory control, and improved neural connectivity could have all culminated in what Isaac's teacher interpreted as confidence. It is confidence that encourages students to give it a go, stick with the question you don't have an answer to, and recognize your progress at the end. In educator speak, this is growth mindset.

The danger of attribution and causation

As a music educator who studies music learning and brain development, I could easily be drawn into the dangerous world of finding what I want to find after reading all this research. It is similar to the experience of breaking your arm and suddenly noticing how many other people also have their arm in a sling but who you never noticed before. Part of training to be a researcher, which is what a Doctor of Philosophy is designed to do, is to be aware of that possibility and remaining as objective as possible about the research.

Psychologist and neuromusical researcher Professor Glenn Schellenberg, who I mentioned earlier, warns very clearly about the dangers of attributing music learning to transferable skills like intelligence and improvements in academic performance. I met with Professor Schellenberg when I visited Toronto and it was a meeting that profoundly affected my own work. I have a note stuck to my computer monitor that reads "What would Glenn say?" to keep me from going too far down the attribution and causation line. I need to balance the lived experience of seeing the impact of music learning on children and schools with my objectivity as a researcher.

Numerous studies have found only small effect sizes attributed to the intervention of a music learning program, which in plain speak means only small changes or improvements in whatever is being measured. Alternative interventions such as playing chess, drama-based learning, and performing, or training and competing in a sport like football have also been studied for their capacity to transfer skills such as improved cognitive function and inhibitory control. Several studies have found that it is the act of taking on a discipline outside of school that improves areas of cognitive development and the activity itself may not matter. These are all important studies to consider and by its very nature the study of human development, which is so individualized and impacted by so many factors, will never provide us with a conclusive answer. Rather, it will continue to help us understand the complexity of the human condition as our world and human beings grow and change.

Looking from many angles

While keeping my objective Glenn hat squarely on my head, it is the breadth and the depth of the research into musical learning that gives me confidence in the findings. We are now heading toward thirty years of neuromusical research, and while in relative terms that is still a young field of research, we have also crested that hill of randomized, longitudinal research with cognitively comparable children. Music and music learning have proven to be extraordinarily useful tools in many fields, including neuroscience, cognitive psychology, developmental psychology, mental health, genetics, audiology, education, rehabilitation, and therapy. Each area of study uses music and music learning to discover different things about their area of interest, but when we look across the field there is a great deal to work with in terms of application in a school setting. Sometimes the research findings can help teachers to understand an event in a new way or they can challenge approaches to learning that we take as gospel.

An important point of difference in the research is the participant group. Many research studies work with typical learners in average schools, and these participant groups help us to learn more about typical human development. Then there are studies that work with participant groups specifically from challenging circumstances. A very well-known one is Harmony Project, which implemented an instrumental music program in gang-affected parts of Los Angeles. Established by Dr. Margaret Martin, an expert in public health, not music learning, the program was designed specifically as a gang-reduction intervention; music learning was just the tool to greater public health outcomes for the community.

However, the impact of this program on auditory processing development and reading acquisition and proficiency has been studied by Professor Nina Kraus and Dr. Jessica Slater, both at Northwestern University at the time of the study. In a nutshell, they found that music learning—in the form of afterschool group music sessions most days of the week, weekend rehearsals, and multilevel developmental ensembles led by expert educators—had the capacity to counteract the negative effects of poverty and trauma because they impacted on literacy development. As Dr. Martin told me in her lounge room in Los Angeles, "Before Grade 2 children are learning how to read, after Grade 2 they are reading to learn. If they miss that window they are in a downward spiral to low self-esteem, low confidence, and joining a gang." For Dr. Martin it was a negative trajectory that she saw the opportunity to change, and Harmony Project has done just that with ninety percent of the high school seniors who participated in free music lessons going on to college, despite the high school dropout rates in the surrounding Los Angeles areas reaching up to fifty percent.

From under my Glenn hat I might say that the research sails too close to causation, meaning the study directly attributes music learning as the mechanism for improved academic achievement through direct transfer of skills. As a researcher I understand and respect this caution, but I also wonder if it misses the broader educational structures wrapped around the music learning that contributed to the success of the program. For example, the explicit focus on self-responsibility and good decision-making as a process in music learning could then have transferred across into the students' other learning and life aspirations. Either way, more than forty percent more students are heading to college after participating in Harmony Project, and while further education is not a silver bullet to lifetime happiness and fulfillment, it also means that a significant number of students may take a different road through life than the one offered to them in the less glitzy parts of L.A.

Confidence makes the world go around

El Sistema is another successful music program, first established in the slums of Venezuela with an aim of combating poverty and trauma. Programs inspired by El Sistema now operate all over the world, taking an afterschool music approach to create community and purpose and to harness music learning as a tool for social change and cognitive development.

I have visited many El Sistema–inspired programs in the United States and Australia and they have produced some fantastic results and rely, as many music programs do, on the generosity of philanthropists and the goodwill of educators for the ultimate benefit of the students. I have also seen music learning programs implemented as part of the school day and school learning experience and I continue to wonder about the outcomes and benefits of each.

Let's go back to Mika's story, which sadly is not the only student's story to have broken my heart but is one that is very memorable to me. The first time I met Mika was in a music class and she was a regular, chatty, lovely ten-year-old. We talked about her instrument, how she had just learned a new note and it sounded like a truck horn, but she felt sure she could make it sound like music with a bit more practice. There was a light behind her eyes and a smile on her face and she bounced out the door at lunchtime.

When I later saw Mika in the classroom I almost didn't recognize her, head lowered and eyes darting around the room constantly. The classroom environment was inviting, her teacher very approachable and her peers were friendly, so by all appearances it was a safe place to learn, but apparently not so for Mika. The interesting thing was that her classroom teacher hadn't yet seen what I had in the music lesson, and when he did get to see a video of Mika in her music lesson he was speechless. How could this nonverbal, traumatized ten-year-old be the same one he taught for the rest of the day?

The answer might be in Dr. Google's definition of confidence, the safety to rely on someone or something again. Music teachers often have this experience. In the world of a child growing up today, a music teacher will often be one of very few adults who are there for a child on a one-to-one or small-group basis every day or week for years. That is a very special place in a child's life, and when that child is recovering from or living in trauma, that person can become even more important. Of course, ongoing music programs are not always available in every school but learning an instrument can be done outside of school. There are a growing number of businesses that are making it possible to lease or rent instruments and music teachers are either promoting themselves more actively or there are local websites where private instrumental teachers can share their details and expertise.

Wrapped up in the bow of music learning is acceptance, challenge, nonverbal communication, reward, bit-by-bit achievement, and often an adult who gets just a little more individual time with them than a busy classroom teacher or stressed parent. It is sad but it is also heartening that this is an option available for the children who may need it most. In Mika's case, music lessons may have been serving a therapeutic purpose more than a cognitive development purpose at first, and this is a pattern I have observed many times in schools. Music programs begin with all the excitement and terrible sounds, and the students may come in with lower reading levels, inhibitory control issues, attention span problems, and motor control issues, so the music learning doesn't seem to move very quickly at first.

In such circumstances music learning can act as music therapy first before then moving into cognitive development.

The first sign that this shift from therapy to educational outcomes is happening is the confidence of the students, confidence to stick at a task, to speak up in class, to try, to fail, and to try again. Often this is the first flag a principal or parent will notice, and while they might not put much stock in the change at first, I can almost guarantee that it will be the first of many more to come, and ones that will change children and schools in ways they can't even imagine.

Further reading

Kraus, N., Hornickel, J., Strait, D.L., Slater, J. & Thompson, E. (2014) "Engagement in community music classes sparks neuroplasticity and language development in children from disadvantaged backgrounds," *Frontiers in Psychology*, 5, p. 1403.

Sala, G. & Gobet, F. (2017) "Does far transfer exist? Negative evidence from chess, music, and working memory training," *Current Directions in Psychological Science*, 26(6), pp. 515–20.

Slater, J., Strait, D.L., Skoe, E., O'Connell, S., Thompson, E. & Kraus, N. (2014) "Longitudinal effects of group music instruction on literacy skills in low-income children," *PLOS One*, 9(11), e113383.

15

WRONG MORE THAN RIGHT
How music teaches persistence

"Tom wants to give up tuba because he isn't enjoying it anymore."

"Aania says she hates the violin, and if she doesn't love it I don't see much point in her continuing."

"Hugo and I have an argument every time I tell him to go and do his piano practice. I just can't see how it is worth upsetting both of us so much if he isn't enjoying it or getting any better."

As a parent, I hear you. I hear you loud and clear. Learning a musical instrument is a wonderful opportunity that many parents want to ensure their child has access to. But it can also become old very quickly and, in truth, the awful sounds and tense exchanges about practice can see that romantic vision of beautiful music drifting out of your child's room evaporate.

There's a certain point in the first episode of *Don't Stop the Music* that made me laugh out loud. The emotional arc of the story had taken everyone through the excitement of hearing the instruments played by professionals for the first time, then the offer of learning one of these instruments and the building expectation that the students could sound just like those from the University of Western Australia Conservatorium of Music Symphony Orchestra. The tempo of the editing along with the rising soundtrack took us to that monumental moment when the first bow was drawn across a violin string ...

Cats screaming. Metal bending. Fingernails down a chalkboard. The sound of a beginner's violin has been described as many things and very few of them are kind or appealing. In the documentary they made use of a single beat of silence before we heard that telltale sound, and every audience I sat with during screenings of episode one laughed at that point.

Of course, the music teachers in the audience knew what was about to happen. They have heard that sound every year, often several times a year, throughout their careers. But they also hear something very different: they hear the potential for that sound to become, over thousands of practice sessions and years of dedication, the beautiful musical sound we are searching for. They know the process intimately and they know the tiny changes in sound that indicate that improvement is happening, little by little. What is required to reach the point where a musical sound and not an animal sound comes out of an instrument is resilience, perseverance, and maintaining the long-term goal of making beautiful music.

Try and try again

Let's go back to Tom's, Aania's, and Hugo's stories. These comments are ones I hear year in, year out. There is often a string of events that have led to a parent approaching me at the end of a concert, parent–teacher interview or via e-mail announcing that their child is not having fun learning music anymore and wants to quit.

Please don't think that I am calling into question anyone's parenting skills or decisions. Raising a child today, whatever you measure that by, is plain hard work. With so many competing opportunities and options to consider while wanting to ensure they have everything they need to enter adulthood and be successful and happy, some things have to give and decisions must be made, and many times those decisions are hard ones. But what I would say is that the struggles, lack of motivation and emotional tensions that Tom, Aania, and Hugo are exhibiting are both real and hiding a deeper difficulty: a lack of comfort with discomfort.

I wish this phrase was mine but I stole it from a visiting expert who spoke at a staff meeting. He was from the field of foreign language and cultural engagement, basically talking about why learning a language other than English and experiencing a foreign culture was so good for a student's overall development. He was eloquent and engaging and, right before the end of his presentation, he said it: experiencing another language and culture is a way to develop resilience and an appreciation of diversity by helping a student to be comfortable with discomfort.

Discomfort with what? Difference? Change? Anything outside of our own experience? All of the above. This got me thinking about how discomfort feels. I have heard people say it makes them itchy and their skin crawls. When I observe a student in discomfort they shift around in their seat, avoid looking at me, or pinch the student next to them to distract them from their own uncertain and uncomfortable feelings. What struck me is that learning is uncomfortable because it puts us into a place of uncertainty where we don't know the answer or have no idea how to start to solve the problem. We will do any manner of things to get out of that uncertainty, to hide the lack of skill or knowledge and allay the fear. Then I thought about the fact that we place students in that uncomfortable position multiple times a day and expect them to rise to it. My question is, do we also give them the skills to be and feel successful at it? Interestingly, research into music learning has found that musically trained students have higher levels of resilience and persistence, and they not only appreciate diversity but seek it out. Are children who are attracted to and seem to revel in the discomfort of learning just built that way or does music learning help them develop those life skills?

A researcher and a parent walk into a bar

There isn't a joke to follow this, sorry, I just liked the heading. For these purposes I am the researcher and the parent, all in one body. One day I was watching my then five-year-old daughter practice her violin in the lounge room. While she played she was crying her eyes out. The researcher in me was excited: here was the experience of discomfort and working through it, no doubt coming out the other end of the practice session with the logical understanding that persistence provided intrinsic rewards and measurable steps toward her larger, long-term goals. That's what all the neuromusical and educational research said, so it must be right.

The parent in me was screaming, "Make it stop!" To see my five-year-old in so much pain was excruciating. Even I, who understood the ear-piercing horror that can be the sound of a beginner's violin, was struggling with the lived experience of the sound she was making. To add such visible frustration to the mix seemed like a bridge too far. But as always, it got me thinking: What is happening inside her brain right now?

Microdosing on discomfort

For those who are not familiar with the term, microdosing is taking a very small amount of a drug in order to test or benefit from its physiological action while minimizing any undesirable side effects. In the case of music learning, the drug is discomfort, frustration, and failure. But only in very tiny amounts, a little bit every day. Think of it this way: when a trumpeter is trying to get a particular note, or any note at all as was the case for one very shy young girl in *Don't Stop the Music*, the act of trying and trying again is a process of microdosing on discomfort. She could very easily have given up, but often it would be her interest and willingness to just keep experimenting that pulled her through. Eventually she blew raspberries on her arm for a week before she could transfer that into the mouthpiece of her trumpet in a way that would yield a loud and triumphant sound. I watched my own mother doing something similar with amazingly complex puzzles, a trial-and-error placement of pieces with a good dose of long-term goal setting to complete the picture in its entirety.

We learn resilience and persistence via many different channels in our lives, whether through observation of how they look or learning how they feel. The last one is particularly important to children because it often doesn't matter what teachers and parents tell us, it is in the moment when we are feeling frustrated and low in confidence that we really learn how we personally express our resilience and persistence. If a child lacks the opportunities to feel those emotions and then work with them, cultivating resilience and persistence is very difficult to do.

Does music learning improve persistence and resilience?

First of all, let's define the difference between persistence and resilience. Persistence is the sticking-with-it moment of continuing with a course of action in spite of it being difficult. Resilience is the ability to recover from difficulties. You can see why I put them together, they are kind of two sides of the difficulty coin. Persistence is keeping going during a difficult time, while resilience is finishing a difficult task (that required persistence) and picking yourself up again. Resilience is also the ability to learn from difficult times in a way where we are willing to both step up again to a challenge and bring new knowledge to the task. For a child this can sound as simple as "What did you learn from your mistake and what will you do differently next time?"

The big question in the neuromusical research field is how much of that persistence and resilience comes from our nature, personality, and genetics, and how much comes from our nurture, own role models, and environment? This was one of the biggest questions to be asked in the research around 2011 to 2014 for the pure fact that if it was mostly nature, then all the research that was being done on musically trained children and their supercharged brains could not be attributed to music learning but was about the children themselves who chose to learn music.

Any activity is easier and more rewarding if you can stick at it. "Time on task" is a phrase used to describe the length of time spent in quality learning of any description, be it learning how to play chess or tennis. It involves a good dose of thinking and an even bigger dose of effort. The effort dose requires both persistence and resilience to keep turning up to the tennis lesson or the chess match every week, even if you miss more than you hit or lose every match. Music learning is exactly the same in many ways, lots of practice, practice, practice. In educational spheres there is the concept of deliberate practice, which is the kind that is purposeful and systematic: in other words, think before you act. It is not rocket science but music teachers spend years and fill countless exercise books after every lesson with notes about what a student should practice over the next week between their lessons. So it begs the question: Does music learning or any other disciplined activity train persistence and resilience?

Neuromusical researchers including Dr. Kathleen Corrigall and Professor Glenn Schellenberg have been trying to get to the bottom of this question for a decade. In a seminal study that looked at the connection between music training, cognition, and personality, they found that when comparing a group of adults and a group of children, all of whom learned music, genes and personality played a big role in the stick-with-it aspect of music learning. They used the Big Five personality dimensions: openness, conscientiousness, agreeableness, extroversion, and neuroticism. Specifically, two out of the Big Five personality dimensions stood out: conscientiousness and openness to experience. This study also looked at many other factors that influence achievement, such as parent education level, socioeconomic status, and general cognitive abilities. They concluded that it was "virtually impossible" to attribute the higher performance on various IQ and psychological tests to the participants' music training.

You might think I would stop writing this book right now, but in fact I find this line of research both fascinating and vitally important to our understanding of the impact of music learning. I use the word "impact" deliberately: impact can be negative, neutral, or positive, and in some cases all three. Impact is the research word for how something changes something else, and with music learning this is nowhere near as straightforward as "music makes you smarter." If anything, this line of research shows us that learning in general is very much an individual process within a highly variable group of influences. Research, particularly scientific research, looks to pull apart a phenomenon to understand it better. Doing so provides invaluable knowledge, but the parts need to be put back together again to understand the individual learning experience for the child that stands in front of us. Music learning may well be a tool to assist in the learning of persistence and resilience, and it is definitely a vehicle to help feel those two skills every time we practice, practice, practice. But like many educational interventions and activities, music learning should not be seen as a magic pill to improve just one or two skills as it has the capacity to enhance multiple skills at the same time. Personality and genetic predispositions may make music learning easier, but is easier better in the long run for our children and students?

I'm not having fun anymore!

Parenting is a tough and relentless gig. While many children get the opportunity to learn music as part of an ordinary school day, others, supported by parents and grandparents, get to experience music outside of school. Sadly, in the market economy model of education that we are currently experiencing, music learning is seen as a luxury, an add-on that might be desirable but not essential. What we may be missing is an understanding of all of the other skills that music learning can teach a child. That is why when a discipline that is teaching persistence and resilience is no longer fun, we as parents seriously consider letting our children give it up.

All of the reasons for quitting given at the start of the chapter have been said to me multiple times. My typical response would be to suggest adjusting practice regimens or changing the music they are learning (often to something technically far easier and from the latest pop charts). One day, without really thinking, I answered, "Okay, sure, let Tom give up his instrument, but what message does that send him?" I was probably just a little too blunt, not my finest hour, but I knew the mother well, had taught her other two children, and Tom was a serial non-stick-with-it student. I liked him and sincerely wanted the very best for his overall development, and continuing to learn tuba was not my primary concern.

His mother took a breath and stared at me for a beat longer than was comfortable. Light dawned behind her eyes as she said, "Oh, that isn't really sending the message I want, is it?" The message she didn't want to send was when things get tough and no longer feel like fun, it is okay to give up. To give you more background to Tom, he was in sixth grade and twelve years old, about to go into junior high school where the stakes were higher and the rope was shorter, and his mother was worried about how his non-stick-with-it approach was going to affect him in the future.

It was the perfect storm for a teachable moment, which for me is one when things seem to be going wrong but the circumstances are in fact revealing an opportunity to learn. The feeling of learning is often uncomfortable, with a good dose of uncertainty and fear lurking around the edges. Try to keep that description in mind when your child comes home from school after being placed in that situation multiple times during that day. No wonder they are a bit quiet sometimes.

In this teachable moment, Tom's mother was ready for a new approach to her son's non-stick-with-it-ness, and it turned out that Tom was, too. This moment opened the door to a plan to switch the expectation of tuba being fun and easy to tuba being the tool to help him experience success in all aspects of his school and personal life. Over the final six months of elementary school he, me, and his mother worked together to set tiny microdosing goals for discomfort, short practice sessions of only five minutes maximum that had to accomplish just one thing. This reverse psychology that he couldn't practice for any more than five minutes made him both think before he practiced and hungry for more time to practice. After just two weeks he negotiated for two extra minutes each day and then out of the blue he told his teacher he wanted to do an exam the following year because he had "cracked the practice problem." I observed some years later that his newfound comfort with discomfort was being deployed as he learned to swim competitively and how to write code. He kept playing and learning tuba until the end of high school and seemed to take in and integrate new information effortlessly. He truly had cracked the practice problem.

Now when a parent or student comes to me and says it is hard and they are not having fun anymore, I say, "Yes! Just think how much your brain is growing and learning right now." Discomfort is a teachable moment and one that music learning throws up all the time—we just need to think of it as an opportunity to learn.

Further reading

Corrigall, K.A., Schellenberg, E.G. & Misura, N.M. (2013) "Music training, cognition, and personality," *Frontiers in Psychology*, 4, p. 222.

Schellenberg, E.G. (2019) "Correlation = causation? Music training, psychology, and neuroscience," *Psychology of Aesthetics, Creativity, and the Arts*.

16

BORN LEADER

How music brings out leadership in children

Have you ever wondered what is happening on a conductor's face during a music performance? Usually all you get to see is their back and their arms waving around, sometimes very dramatically. As an audience member you might try to put together a story from this of what the conductor might be asking of the ensemble. But would you like to know what the look on their face really is?

It can change quite dramatically. Overwhelming confidence. Deep love. Manic anger. Sometimes unadulterated primal fear. It is an act worthy of an Oscar for nonverbal communication, and if you ever get the chance to sit on the other side of the conductor and watch their act, take it. It will change your view of musical performance forever.

The conductor is the obvious leader of an orchestra or band. But there are many different ways of being a leader and leading, and these don't always involve being at the front and the center of attention. A leader can also lead from the middle as a role model or from behind as a motivator. Most of my students will never get the opportunity to lead from the front, but they may want to take up the opportunity to lead in the many others ways that they can.

This chapter is about how music learning helps find the born leader in every child. The act of performance, whether it be in a group or as a solo performer, is a tool to both find and teach leadership in every child. This is not necessarily about being the type of leader who leads from the front; we can only have so many of those in the world. Finding the inner or born leader in every child often means knowing when to lead from the front, when to follow another leader, and, possibly most importantly, how to lead in many different and probably not so obvious ways. Music learning is a vehicle to help children learn how to contribute as a productive member of any team or society.

Who is the leader?

I want to share the story of one of my most epic musical fails in recent years. Picture this: 180 students from the age of twelve to eighteen years on stage with brass, wind, and percussion instruments. All eyes are focused on me and their performance-ready bodies are coursing with adrenaline and a healthy dose of excitement and fear mixed together. There is me up the front as conductor, oozing confidence and visually preparing the audience to be wowed by the might and precision of this awesome ensemble. Then I motion the downbeat with my baton—that's the first beat played—and it all sinks into quicksand.

To be fair to my wonderful students, this is not strictly true. The whole performance did not sink into the abyss but one section of it definitely did not go to plan, and in that moment, a moment where no one can speak or deliver instructions, this 180-strong group of teenagers made a vital team decision.

The piece of music had a driving snare drum part underneath a floating, noble melody. Think of the snare drum part as something you would hear in a marching band, a repeated pattern that feels like it drives the music along while remaining rock-steady, almost like a robot. It had been practiced in individual ensembles before these three concert bands came together, and the sense of a shared pulse was a big part of what we had concentrated on. I often get my students to put their right hand across their body and over their heart while they try to keep a shared beat for as long as they can. It is such a primal experience, like dancing around the tribal fire, and one we don't have the opportunity to experience very often in adult society. During our rehearsal period we had worked very hard to share this pulse and feel like one strong, synchronized heart.

So when one young drummer lost that heartbeat and went rogue, driven by adrenaline and nerves and the weight of expectation, the whole piece went sideways very quickly. His repeated rhythmic pattern started too fast and just kept getting faster, like a snowball rolling down a hill. When something goes astray, musicians are trained to focus on the conductor to find the shared beat again and adhere to their nonverbal signals for the greater good of the team and the performance. And that day, 179 pairs of eyes did just that, asking me with the blazing whites of their eyes what they should do.

I turned my energy to the young percussionist, trusting the other 179 pairs of eyes to follow me, and basically did a "whoa there" thing with my hands and body, revealing none of the primal fear and mild rage I was feeling. The look on my face was "we got this," we just had to pull it back and the train would get back on the tracks.

The drummer looked at me and his blood-drained face said, "I can't stop! I am trying to tell my arms to slow down and find the beat but they aren't listening. What do I do?" It is distressing to watch someone so out of control, but meanwhile, the music is still going, the audience is still watching, and by now the musically experienced parents might be pulling that face that asks "Is something going wrong up there?"

Here comes the moment of deeply human teamwork and trust. I looked back at 179 pairs of eyes and we all came to the same conclusion at the same time: our young percussionist is the leader now and we need to follow him. We all knew it wasn't going to make for a sensational performance, but we also knew that the larger picture of success was dealing with the present, a situation that had never once happened in rehearsal. The section of the piece went lightning fast and we had a triumphant final note together, but because we were going so fast not everyone reached that note together. It sounded like a car crash at the end, a big bang followed by little bits of bumper falling off and maybe a wheel rolling past before it lost momentum and toppled over. We got there, but it was not pretty.

As the most obvious leader up the front, I could have been disappointed in the young percussionist, distressed about all the seemingly wasted rehearsal time and embarrassed for not giving the audience a performance that was seamless and polished. We finished the piece and I did one of those bows to the audience that is so infused with confidence and a "we nailed it" attitude that the parents might have second-guessed themselves and wondered if they were just hearing things. Never show them you got it wrong is a mantra for all performers.

Typically, a conductor would leave the stage first and then the performers would follow, but because we were such a big group and you don't want bent trombone slides and wrecked clarinet reeds, the students filed off first. The young percussionist was at the back, his pale face now replaced by skin the color of a tomato. Unbeknown to me, who was still on stage, many of the students were giving him high fives, shoulder shakes and comments like "If you are going to get it wrong, get it *really* wrong." As one student said, "You failed good," and that really sums it up: the young percussionist went for it and he missed by a mile, and that is great.

You failed good

Music performance is a high-stakes and very public opportunity for great success and awesome failure. It is a moment in time when hours of rehearsals built on years of development come together to show themselves. It is so easy to fail and so hard to succeed. There used to be an ad campaign about driving safely that simply said "Strange things happen on country roads." I often think of that in music performance; things that have never happened in rehearsal and you have no way to prepare for them to happen in performance. The resulting adrenaline heightens awareness, which is great for performance, but can also speed up your heartbeat and shorten your attention span, leading you to lose the ability to focus on the moment and become overwhelmed. When it happens, the team has to adjust, problem-solve, weigh up options and go with one, all without a single word being spoken—these are intuitive, in-the-moment decisions informed by hours and often years of team-building.

There is another mantra in education about learning through failure. A student in *Don't Stop the Music* said it best while in conversation with Australian jazz musician James Morrison: "I love failing because I know that I am learning." But while failure is spoken about as a thing we should all do often, actually doing it is not so nice. Failing is shooting for the stars and missing, failing is trying what you believe is your hardest and not getting the result you want, and failing often feels horrible. As humans, we try everything we can to avoid failure because it just doesn't feel good. It takes a lot for us to not only learn from failure but to even seek it out in the first place.

Music performance is the potential for failure amplified up to ten out of ten. It is public, often in front of parents who we like to impress (both myself and the students) and sometimes in front of school leaders or even assessors. There is always the potential for failure and judgment, but what is interesting is that as this is often done as a group, that makes it just a little less scary. Built into the music performance process is the ability to both experience failure and learn from it, and to experience it often and publicly. The frequency is the key to building skills individually and collectively, because the more an ensemble both succeeds and fails, the more the members learn personally and as a team.

Teamwork through our ears

The roles of leader and follower in a music ensemble are fluid concepts. One moment the trumpets might be the leaders with the melody, but eight bars later they need to become the followers when they have the accompanying parts as the flutes take over the melody. The idea goes across all music ensembles: the violins in an orchestra or the sopranos in a choir might switch from melody leader to accompaniment follower at any given moment. Then there are the times when everyone is playing or singing together and collectively creating the sound, but their layers have corresponding roles.

The exchange of roles happens both vertically, through the sound in a given moment, and horizontally, as the piece of music progresses. Professor Stefan Koelsch, one of my favorite neuromusical researchers, whom I got to visit on the only sunny day they had in Bergen, Norway, in April 2016, continues to use music to reveal how the brain works. One of his most useful pieces of research for me is a model showing how the brain extracts information from music sound, analyzes it, compares it with other music, puts it all back together and then responds to it while also extracting meaning and emotion from it. This process occurs incredibly quickly and, as I described in Chapter 1, looks like a three-dimensional squishy ball of cognitive activity in the brain that is processing music both in the moment and over time. The amazing part is that this is done without high levels of conscious thought—it just happens as a part of our supercharged auditory processing.

Add to this incredible but unconscious auditory processing the highly conscious and heavy cognitive task of figuring out where your part, at any given time, fits into the bigger picture of the music. Am I melody or accompaniment? Do I need to fit my sound inside someone else's sound (a tricky auditory concept to first understand and then do)? If the tempo or speed of the music is getting slower or dragging, can I change that if I play a different way? If I have the best sound in my section, why does my teacher keep asking me to blend voices, and why should I?

This last one is particularly interesting with children who by nature want to stand out in a musical performance, bound up with their concepts of leadership and self-worth. The music rehearsal and performance environment is a perfect place to help children form multiple ideas of what contribution might look like to the musical team. This can recruit both auditory processing networks as well as the larger group of executive functions, one of which is the skill of resolving cognitive and emotional conflict. On a day-to-day basis, this can be the desire to say something honest and forthright to someone, an emotional response, and then to reconsider it based on the awareness that it might not get the reaction or response you are ultimately looking for, a cognitive response. Put into very simple terms, it's the "think before you speak" mechanism that many children need to work hard to develop during childhood. In a music learning setting, this could be wanting to play a section of music really loudly on your trumpet because you like it but choosing not to because it overshadows the other parts or distorts the musicality of the ensemble.

Other executive functions get a workout in the music rehearsal and performance setting. There's the ability to plan and use spatial skills, for example, working out when you need to raise your instrument or take a breath in order to play at the right time. This might sound easy, but for a ten-year-old to look ahead in the music, to continue counting their bars of rest, to remember how long it takes to raise their instrument and prepare their body to play it is a monumental cognitive feat. If they get it wrong they run the risk of knocking their teeth out or making a very unattractive sound with their instrument (I have seen both happen, often).

One of the most interesting things to me about music rehearsal and performance as a learning experience is how similar it is in many ways to the concepts we talk about in sports, like training, games, skill-building, and teamwork development. I have spoken to many children and adults who can directly connect what they learn in sports and music programs to their understanding of and development as a leader. If I put a bunch of physical education and music education teachers in a room together with the fertile question of whether their own experience develops children similarly or differently, there is a good chance they would find common ground.

I have been lucky to work with an inspiring researcher named Professor Phillip Tomporowski, who leads the Physical Activity and Learning Program at the University of Georgia. We got talking about common processes across sports and the performing arts, including practice regimens, rhythmic coordination, and multisensory learning. What struck me in our conversations was the fact that both our fields were developing and enhancing cognitive bodily processes in a team setting. The difference is that sports are by their nature competitive, whether you're racing a clock, a distance, a team, or ourselves. Music performance can also be competitive in an exam of eisteddfod format but in a regular end-of-term concert there is no win, just success of our own measuring. To unearth and enhance leadership in children that will serve them well throughout their lives, they need to both be able to define a win by outside measures and appreciate that win from within themselves. Nature *and* nurture, sports *and* music, leader *and* follower.

Further reading

Collins, A. (2013) "Neuroscience meets music education: Exploring the implications of neural processing models on music education practice," *International Journal of Music Education*, 31(2), pp. 217–31.

Koelsch, S. (2011) "Toward a neural basis of music perception: A review and updated model," *Frontiers in Psychology*, 2, p. 110.

Tomporowski, P.D. & Pesce, C. (2019). "Exercise, sports, and performance arts benefit cognition via a common process," *Psychological Bulletin*, 145(9), 929–51. doi: 10.1037/bul0000200.

Trainor, L.J., Shahin, A.J. & Roberts, L.E. (2009) "Understanding the benefits of musical training," *The Neurosciences and Music III: Disorders and Plasticity*, 1169, p. 133.

17

I DO, I UNDERSTAND

How music supercharges your child's memory

"Hang on, let me write that down."

It is quite likely that if you walk into a school you will see this little saying on a poster or computer screensaver: "I hear and I forget, I see and I remember, I do and I understand." Often attributed to Confucius, it is commonly used in educational practices as a reminder that to learn something, students, be they a child or an adult, need to interact in a practical way with the thing they are learning about.

When we do something we understand it, which is the reason we don't just learn science concepts from a textbook anymore. Not only that, by doing something we commit that understanding to our memory banks and use it again later. Music-trained children have been found to have exceptionally agile memory systems because learning music wires the brain for effective storage and retrieval of memory. Having a good memory is the basis of daily school and life experiences: "Do you remember the experiment we did yesterday? Now let's build on that understanding today." As tweens enter adolescence and junior high school their memory system really gets a workout, and in this chapter we look at how music learning can build the pathways along which that system can thrive.

Remember these twenty things

Some years ago I was in a professional development workshop with about forty high school teachers, all highly experienced in their fields of anything from physics to physical education. I was the only music teacher.

The facilitator, an expert in learning who would become one of my foundational mentors, named Dr. Julia Atkin, set the group a seemingly simple task: memorize these twenty regular items in order in two minutes. I say "seemingly" because it wasn't simple at all, it just sounded it at first blush. She added two caveats: you couldn't write anything down and you couldn't say the list out loud.

Here is an example of a similar list:

1. Lamp
2. Surfboard
3. TV
4. Coffee
5. Computer mouse
6. Chair
7. Wine

8. Tree
9. Lightbulb
10. Dog
11. Fire
12. Water
13. Cake
14. Plane
15. Tooth
16. Hat
17. Apple
18. Bicycle
19. Ball
20. Piano

The two minutes started and the participants assumed all the positions I see with my students when they need to concentrate on a task: head down over the table with a scowl, sitting back very quietly mouthing the items in order or staring into space then looking back at the list when they got stuck.

Two minutes was up. The next instruction was to write down the list in order. Do you want to guess how many people remembered all twenty items on the list?

Only one person got all twenty items correct and in the correct order. Me.

Now before you jump to the conclusion that my neuroscientifically interesting brain is awesome, that's not why I am sharing this story. I am sharing it because it was the most powerful object lesson I have had in how differently we learn and create memories, in this case a specific kind using our working memory to temporarily store information for quick retrieval.

I will never forget the gasp in the room. Here I was, an early career teacher with only three years of practice under my belt in a room with educational experts and school leaders averaging more than twenty years of experience. What's more, I was a music teacher, and in the scheme of physics, math, and English that was perceived to be a fluffy subject that doesn't work on right and wrong answers, just subjective judgment.

Dr. Atkin asked a few other teachers how they went about memorizing the list. Some connected each item with a saying, some memorized the list by sight and a few made up a poem using the words. Most of the teachers got to seven or eight items, a small number to just over ten, and the highest apart from me was sixteen. Dr. Atkin then shared that she had done this exercise with more than fifty groups and each time only one person had got to all twenty items, and that person was more than likely to be a music or arts teacher. Dr. Atkin then turned to me and asked, "So how did you do it?"

I told them that I made a four-dimensional film of the list in my head, which meant I could see, smell, hear, and feel the story I created. I figured you could find most of the items in a house, albeit a slightly weird house as the list went on. I walked through the creaky front door to see a tasseled art deco lamp. There was a slight whiff of cigarette smoke in the air, which made me feel a bit sick. I turned to the right and saw a dusty surfboard leaning in the corner that reminded me of one in our garage that my husband purchased with great enthusiasm on a south coast trip and has never used again. Next to that was a TV; the sound was down but there was an advertisement playing on it with coffee beans pouring across the screen.

Do you get the idea? As I write I realize that the 4-D film takes forever to describe but a moment to experience. Some two decades later I can still remember the first six items on the list and how I incorporated them into my film.

Around this room of experienced teachers and leaders from a myriad of different teaching disciplines such as history, science, and physical education there were smiles and heads shaking in disbelief along with spontaneous applause at the end of my description. People commented, "How did you think of doing that?" and "That's so creative," but I still remember Dr. Atkin setting the group straight pretty quickly.

She said something along the lines of: "She probably didn't 'think' of it, and yes, it is creative, but it is also more than that—her brain is wired to create memories and learn much more quickly and accurately than yours. Being a musician has set her up for that."

This learning development workshop was in the very late 1990s and Julia was hinting at the new research just emerging that musically trained adults appeared to have better memory creation, storage, and retrieval systems. Since then, researchers all over the world have studied musicians to understand how memory works, how memories get lost or rewritten, and how music can help to heal our memory systems after damage and trauma.

The memory-making conundrum

In the early 2000s, a team led by Dr. Lorna Jakobson explored the previously researched phenomenon of musically trained adults having improved skills in the recall of lists and prose. They found that these specific recall skills may be related to enhanced verbal memory, an enhancement which was attributed to music learning. Why is enhanced verbal memory a big deal when we have so many ways of recording and transmitting information now? Is remembering what we hear so important these days? Step into any school and you will immediately see how important verbal memory is: most instructions are given verbally, educational practice favors discussions and question-and-answer techniques, which are predominantly verbal, and social interactions in the classroom and on the playground are all verbal. Schools are very talky places.

There is a world of difference between hearing a verbal fact or instruction and understanding what you have heard. Understanding can mean anything from conceptualizing the information, connecting the information with other information you have already understood, matching the information or finding it doesn't match or challenges your previous understanding—the list is endless. It kind of falls into "if a tree falls in the forest" territory when we ask, "If a child hears a fact or instruction verbally but doesn't do anything with it, did they learn anything?"

Verbal memory isn't just one stand-alone part of the brain, it is just the beginning of a fascinating journey of processing, memory-making, and learning. These operations were thought to have a significant impact on academic performance since the ability to hear something once, process it, then add it to your brain library of knowledge and skills with multiple tags and associations for easy retrieval is a good description of a fast and agile learner to me. Musically trained adults and children seem to possess these skills, so researchers were searching for both an understanding of how these processes occurred and if/how music learning enhanced these processes.

The answers are still being revealed and, I suspect, will continue to be discovered for many decades to come. One of the first research findings in this field was improved verbal memory, which is the ability to remember what you have heard in a way that you can recall it for later use. On a daily basis we do this all day long and utilize our working memory— our temporary folder of things to keep in mind for the next minute, hour, or day—to store that information. Children's working memory folders are smaller, often hold less, and they will misplace their contents if they are tired, hungry, or distracted. However, in the natural development of a student this folder gets larger and more reliable. Around the ages of ten to twelve years teachers can often identify students who are going to struggle academically because they are already struggling with their working memory and memory retention and retrieval.

These struggles could come from issues with auditory temporal-order processing, which is the skill of first hearing then understanding in relation to time the information they have heard. As I have said, the mechanisms the brain uses to learn are nowhere near as straightforward as a series of steps—the brain is engaging multiple areas and functions both simultaneously and in response to different processes. It is not as simple as "just listen" or "repeat back to me" when it comes to improving auditory processing of information.

As researchers continue to investigate this aspect of brain development, the study of musically trained adults and children has helped enormously. In their recent chapter in the huge

Oxford Handbook of Music and the Brain published in 2018, Dr. Psyche Loui and Rachel Guetta reviewed the connections between music, attention, executive function, and creativity. You might wonder what that has to do with memory but the making, storing, and retrieving of information in our memory requires us to pay and maintain attention, using executive functions such as auditory processing at the same time as our sensory processing in order to not only learn something but then do something with that information, which in its broadest sense is to consolidate and create.

Loui and Guetta expertly take the reader on a journey of what we know, what we think we know, and what we aren't sure that we know about the connection between these important brain functions and music learning. We do know that music learning enhances auditory processing. This is a reasonably straight connection as we know that music learning forces activity and connectivity within the brain that allows it to perceive the nuance, meaning, and changes in sound, which can be transferred across to language. We think we know that music learning helps to extend attention stamina and modulation. This can happen through practice on an instrument but also learning and performing in an ensemble, a social group-learning experience that tends to give children a reason to keep paying attention after they would have normally become distracted. We think we know that the nature of music learning engages most of our senses and also requires them to work together and process sensorimotor information rapidly and in concert with the cognitive and perceptual informa-tion we are also processing. Loui and Guetta explain from multiple angles that both the parts and the whole experience of music learning appear to have an impact on enhancing either the individual systems or the connection between all of the systems that help us learn. What we don't know yet, as always, is how much and in what way the nature aspects of genetics and personality impact on the nurture aspects of experience in school and home and access to music learning.

I did it my way

Learning how brains process information into memories made me look back on my experience with the list of twenty things very differently. Without thinking anything of it, I tagged the list with sensory information that took it from a meaningless series of items to a meaningful and individual work of art that I can still experience.

How do I think my music learning enhanced that skill? Probably by making me hypersensitive to any and all information that my brain and body were taking in, and then integrating that information into both a cognitive and perceptual understanding. Indeed, back in 2003 Dr. Lorna Jakobson and her team suggested the next steps in the research should be to explore the "cognitive/perceptual process that is directly strengthened by music instruction." They and others are still working on it, and how humans process information and learn is pushing our thinking with every new experiment.

We have gone on a windy road in this chapter and you might be struggling to connect it all back to your own tween, your class of tweens, or what you were like when you were a tween. Music learning at this age can be pushing the limits of your tween's available time and engagement levels but it may be preparing them for their next cognitive test. In a nutshell, the further along the cognitively connected track your tween is the better, because what comes next will test, change, rearrange, and rewire your tween's cognitive capacity: puberty.

Further reading

Besson, M., Chobert, J. & Marie, C. (2011) "Transfer of training between music and speech: Common processing, attention, and memory," *Frontiers in Psychology*, 2, p. 94.

Degé, F., Wehrum, S., Stark, R. & Schwarzer, G. (2011) "The influence of two years of school music training in secondary school on visual and auditory memory," *European Journal of Developmental Psychology*, 8(5), pp. 608–23.

Jakobson, L.S., Cuddy, L.L. & Kilgour, A.R. (2003) "Time tagging: A key to musicians' superior memory," *Music Perception: An Interdisciplinary Journal*, 20(3), pp. 307–13.

Loui, P. & Guetta, R. (2018) "Music and attention, executive function, and creativity" in M. H. Thaut and D. A. Hodges (eds), *The Oxford Handbook of Music and the Brain*, Oxford: Oxford Handbooks.

PART FOUR

ALMOST A GROWN-UP
12–14 YEARS TO ADULT

18

MANAGING IT ALL

How music helps children with balance

"He's playing basketball, doing really well at school, and playing cello in the top orchestra at school—he just seems to thrive on doing it all." I overheard a proud parent talking to a friend outside the supermarket. Having a son who is managing and indeed thriving at school is certainly something worth sharing, and often such a description is fueled by pride mixed with relief. These teenagers will more than likely hold a leadership position within the school, volunteer to help out someone who needs a hand, and be generally kind and decent humans. They are teenagers who just seem to have got it together.

Another layer of complexity to remember is that while these teenagers are keeping it together, their bodies are going through the significant changes associated with moving from childhood to adulthood. Puberty has the potential to impact on their memory, emotional stability, impulse control, attention skills, and a whole range of other executive function skills that directly affect learning.

As a junior high and high school music teacher I have the privilege of teaching my students for up to six years. I get to witness and influence their growth, both personally and academically, across the eventful and formative years between the ages of eleven/twelve and seventeen/eighteen. Not just some but most of my students are the ones who seem to have got it all together and I have been able to observe how I believe music learning contributed, taught, and motivated them to do so. In this chapter I want to share two stories, archetypal but drawn from real life, about how music learning seemed to influence these students' growth and how the research might begin to explain why this happened.

Daniel

Daniel had the gift of gab, meaning he could talk his way into or out of anything. He could read people very well and expertly manipulate them even at the age of thirteen. His skills at manipulation were both frightening and intriguing to watch; he was on the road to a very successful career as either a spy or a con man. He also couldn't maintain focus on his schoolwork, possibly because he found it profoundly boring, and the most common phrase on his school report was along the lines of "Daniel has significant potential to achieve well in school but lacks the focus and discipline to complete the tasks set for him."

Daniel was on a path to trouble: trouble in school, trouble with his parents, and one day possibly trouble with the law.

He was thirteen years old, just a boy. Was he just testing the boundaries or was intervention needed? After a string of increasingly severe infringements, Daniel and his room tutor, who was in charge of his pastoral care, turned up at the music classroom door. "Send him to music, they'll fix him" was how I would paraphrase the conversation that led to this point. The tutor didn't know what else to do but he had seen similar boys "come good" when they entered the music program.

Now, I could launch into a heartwarming story of growth and acceptance along the lines of *Mr. Holland's Opus* or an episode of *Glee*, but that isn't what interests me in these types of situations. What interests me is why—why on earth would being involved in music "fix" Daniel? What is it about music and the learning environment and strategies that are used that would help him "come good"? And it is also worth asking, would it be the same process and result if he was sent to join the basketball team or doing community service at a soup kitchen?

Here is what happened. Daniel was a smart student with sloppy impulse control. These students are always interesting to me. On the surface they seem to distract every other student in the class, which means the teacher needs to deal with them to keep the entire class on track, but under the surface this incessant need to cause trouble may be fueled by anything from a lack of understanding of the content being taught, to a need to hide dyslexia, or another language processing disorder, to a behavioral disorder like ADHD, through to the content just being too easy for the student. I often think of it as the behavior being the pain in your lower back which is actually stemming from an issue in your feet.

In Daniel's case I think it was a bit of boredom with the content mixed with inhibitory control issues, possibly influenced by undiagnosed ADHD. The music learning prescription was easy: teach him how to play drums. Here is why it worked.

Learning to play drums, which is made up of increasingly complex rhythm patterns put together, takes discipline and intense focus. A student starts with just four beats and then learns all the combinations in just four beats, which is the equivalent of about two and a half seconds of concentration. The aim is to play each pattern so correctly, so metronomically, that they are like a robot, and then they have to do it again. It is deceptively hard to do. For Daniel this was a tick in the engagement box: he had finally come across something that was hard for him to do and he couldn't talk his way out of it. Indeed, it was the first thing he literally couldn't do while he was talking. Soon Daniel moved from playing single patterns on one drum, just using his hands and arms, to playing two, three, or four patterns using all of his limbs. And then he had to deliver these steady, reliable rhythms in a band environment where everyone was counting on him to get it right. No band can perform with a flaky, unreliable drummer.

Learning drums also meant remedying the possible neural processing issues he had from his undiagnosed ADHD. By studying musically trained children we have come to understand that ADHD is in part a mistiming between the auditory, motor, and visual cortices; the faster and more synchronized timing between these three areas which was evident in musicians was one of the researchers' first discoveries. Focusing on rhythm accuracy and repetition may assist in realigning and strengthening that timing, as shown in a number of studies when used as therapeutic intervention with children and adults with ADHD. Drums were also helping him work with his inhibitory control—learning to play an instrument was also about learning what it felt like to get it wrong and try again. What lives in the middle of those two steps—getting it wrong, trying again—is the feeling of frustration and the slow building of resilience, as well as a side order of attention stamina and creative problem-solving.

Music performance is ephemeral, meaning it only exists in the moment that we hear it, and therefore while Daniel might be able to play a rhythm pattern in a lesson, it also needed to be re-created on stage during a performance, with all the nerves and excitement that come with it. This heightened experience, the need to get it right mixed with the thrill of performance, probably activated Daniel's brain reward network. He got the same (or an even larger) buzz from performing music under heightened conditions that he got from breaking a school rule or successfully swindling another student out of something. Music learning wasn't teaching him to reject the thrill he was beginning to enjoy from his particular set of skills, it was replacing it with one that was internally generated, possibly more addictive, and required social interaction and personal responsibility to achieve.

Music learning did indeed "fix" Daniel, but sometimes I think of those drumsticks as lightsabers that just helped him choose the light side instead of the dark side in applying his particular skills. By the age of eighteen Daniel was a leading student at the school and much loved by the younger musicians, who looked up to him as a role model with high standards of personal responsibility. His impulse control issues will always be with him but he has had the opportunity to work with them, become aware of their influence, and test out his own strategies to manage them throughout the rest of his life. He doesn't play drums as an adult but what the experience taught him has molded his cognitive and emotional processes to change his life path.

Talia

Talia could do it all. Captain of the debating team, academic excellence in everything, violinist in the state symphony youth orchestra, and tutor to younger students in mathematics whose parents couldn't afford to pay for it. She gave and excelled everywhere.

Believe it or not, these are the sorts of students that I often get most concerned about. It is almost like they have found a list of things perfect students should do and they are just ticking them off. At some stage, often in the first years of high school around fourteen or fifteen years old, the balance of perfect will become unbalanced, unmanageable, and unsustainable. This is when the goody-two-shoes morphs into the anxious perfectionist, and the puberty monster just adds to the roller coaster. Within the multiple skills that make up our executive functions, the balance can definitely go off.

Tempting as it is, I am not going to launch into a rendition of "Let It Go" here to explain how Talia found her true power and let all the social and academic expectations fall away. This story is about how she used music to get her executive function groove back.

Learning violin often starts at a very young age because

it honestly just takes longer to learn. My primary instrument was clarinet, and when you want to play a D you put down the same fingers every time and get the same note. Three years ago I took up cello and that was a whole new world of learning pain. There are no valves or finger holes, just a fretboard where you need to put your finger down in the same place every time. Think of it as aiming to get a bull's-eye in darts a hundred times in a row.

Talia could play violin very well, but to do that she had to follow all the rules, all the time—correct bow hold, correct bow pressure, relaxed shoulders, left wrist angle, the list goes on and on. Luckily, when you learn to play an instrument, a bit like driving a car, many of these things become second nature and Talia didn't have to think about them. But she loved rules and lists and ticking things off. In a very different way, these sorts of experiences were the ones that got Talia's reward network pumping, but eventually the lists and rules were no longer bolstering her effectiveness and learning, they were starting to strangle them.

It's music class, the period before lunch on a Thursday. Jazz is the topic and improvisation is the musical form we are learning about. It's a way of making music where a group of players agree on a chord structure—let's say eight or twelve chords in a row that work well in a sequence—and then each person takes a turn making up a melody and playing it over the top of the chord structure the others keep. That's a simple description, but it is incredibly complex and personalized; there are thousands of rules that a jazz musician can choose to follow or break to see what type of art pops out. Sometimes it works and sometimes it doesn't. Imagine you and a friend were given the same canvas and the same paints and were told to paint the same person's face. Both paintings would be very different because you make thousands of different decisions along the way.

Back to improvisation class. Talia had her violin ready and was very eager to get it right. I love this moment with teenagers who may have been indoctrinated into the "right" culture of schooling rather than the "learning" culture. "Well, there could be a right improvisation, but who determines right, and what are the rules of right, and is your right the same as my right?" Perfectionist teenage mind blown.

Talia was not comfortable with this at all and proceeded to do all the things students do to get it right—take over the group, set out the rules, tell other students when they get it wrong as per her right rules. As a teacher it is like watching a train wreck in real time, but it is such an important train wreck for everyone involved because it is one of those precious teachable moments.

Talia's issue, with both jazz improvisation and most of her life motivations and choices to this point, was that she could only see one answer to any given problem. She had survived and indeed thrived in the eyes of her teachers and peers by finding the singular, right answer. Now in her early teens, when learning and life were getting more complex, the big question for her was what do you do after the first "right" answer doesn't work? Talia had several parts of her executive function skills out of balance: her problem-solving skills didn't extend to multiple solutions and she did everything she could to avoid discomfort, and discomfort for her was not knowing what the "right" answer was.

Jazz improvisation pushes all of those buttons because ordinarily you can't improvise alone and you need to find solutions in a socially motivated setting. Yes, Talia tried storming off and then taking over, but in the end she had to embrace her "right" answer at the time, in that moment. One of the first definitions of executive function I ever found was the ability to resolve cognitive and emotional conflict, and to me this is what I saw Talia learning in that music class.

Skip forward to her final exams for high school and there is Talia on stage, improvising on her violin with a jazz quintet. She is still a student who seems to do it all and more, but the emotional stability that had become so unbalanced when she was fourteen is now rock-steady. True, this could just be down to maturation and experience, but Talia herself can express what jazz improvisation has taught her: "that the right answer is the one I have now and that there might be a different right answer tomorrow." She says it with an offhandedness that I always get caught out by, and I notice a calmness and mindfulness in her that I never would have predicted she would grow into when she was fourteen. Research into jazz improvisation shows that musicians' brains become calmer and use less conscious thought in order to access their social interaction and empathy, possibly the source of our creativity. Might this have helped rebalance or expand Talia's one-solution processing? She had found her executive function groove in the most interesting of ways.

Decision-making skills in teenagers

It sounds like an oxymoron: we can so easily assume that teenagers have no decision-making skills because we as adults perceive that they are making the "wrong" decisions. As a teacher and researcher, it is how these decision-making skills get developed and the environments, processes, and serendipitous events which shape our decision-making paradigms that interest me. Early teenagerhood is when children are getting a glimpse, physically, cognitively, and emotionally, of their adult selves, and this is where the executive function skills that inform the decisions we make for ourselves and others really start to develop. In Daniel's case, parts of his skill set, his inhibitory control and his attention, were not firing effectively, and music learning may have helped to improve that communication. In Talia's case, her problem-solving skills and cognitive/emotional conflict were heading toward stunted overdrive, and music learning may have helped her expand and calm those processes. Daniel came to see his impact on others through becoming more disciplined, engaged, and aware of the world outside of himself, while Talia came to expand the possibilities for herself through becoming more mindful, open, and adaptable.

I am aware of a few gender stereotypes in these stories: the boy who plays drums and can't pay attention and the girl who plays violin and is a perfectionist. I have observed these traits in boys and girls, percussionists and string players, and in all other combinations. They are archetypal stories based on a number of very memorable students, and what they do achieve within the archetype is a reflection of what the research is beginning to explore. So much of the research focuses on the younger years of development because this supports the researchers' primary goal of understanding human brain development. A smaller number of studies look at the later years of childhood, and one in particular asks the question you may also be asking: What happens if you start learning music after the age of seven or eight?

Not all is lost, and in fact, something very important is gained. A 2017 study led by Dr. Kirsten Smayda compared adults who did no music with those who started learning music before and after the age of eight and looked at a number of their executive function skills around social decision-making. This preliminary study found that the "later an individual begins playing music in childhood, the greater the benefit to their decision-making skills as young adults." We are yet to investigate why this might be so or how it might impact in later life, but this finding made me think about what is involved in making a decision: balancing cognitive and emotional conflict, balancing the needs of others with our own needs, balancing what we feel like now versus what we might feel like later, and judging time. This can extend from decisions about friends and relationships to how long an assignment will take and what to do with competing academic priorities. We make decisions every minute of every day, and the more skills and experiences we give our young teenagers in this, the better.

Although I used archetypes for my two stories, I can tell you where Daniel and Talia have channeled their balanced executive function skills. Daniel is in sales, and as the saying goes, he could sell ice to an Inuit, but only if they really need it. Talia is a chief resident doctor in a large hospital emergency room. She loves the rules and the structure, but she loves it even more when it all goes wrong and she gets to improvise.

Further reading

Dhakal, K., Norgaard, M., Adhikari, B.M., Yun, K.S. & Dhamala, M. (2019) "Higher node activity with less functional connectivity during musical improvisation," *Brain Connectivity*, 9(3), pp. 296–309.

Puyjarinet, F., Bégel, V., Lopez, R., Dellacherie, D. & Dalla Bella, S. (2017) "Children and adults with Attention-Deficit/Hyperactivity Disorder cannot move to the beat," *Scientific Reports*, 7(1), p. 11550.

Seither-Preisler, A., Parncutt, R. & Schneider, P. (2014) "Size and synchronization of auditory cortex promotes musical, literacy, and attentional skills in children," *Journal of Neuroscience*, 34(33), pp. 10937–49.

Smayda, K.E., Worthy, D.A. & Chandrasekaran, B. (2018) "Better late than never (or early): Music training in late childhood is associated with enhanced decision-making," *Psychology of Music*, 46(5), pp. 734–48.

19

SEEKING OUT THE NEW

How music teaches children to create and innovate

Children just love Waldo. If you have never had the joy of "reading" a *Where's Waldo?* book, this is how it works. Imagine a double-page large-book spread of the most intricately detailed drawing of a scene, maybe a beach or a forest village. There are literally thousands of tiny cartoon figures doing all sorts of sometimes quite crazy things. The point of the drawing is to find Waldo, who is dressed in a red and white striped top and bright blue pants with a white and red beanie on his head.

To many adults this might sound like a torturous waste of your time, but to a young child it can be the most enthralling thing they do that day. It is the challenge of finding just the right combination of clothing and character among all the other distractions, the joy of seeking out difference and the odd one out.

I see this type of brainteaser everywhere, from two cartoons side by side that have eight subtle differences through to the cladding on a new building where all the panels are grey except for a few randomly scattered, brightly colored ones. As humans we like and search out patterns, they make us feel comfortable in our surroundings, but as soon as we identify a pattern we seek out breaks in it to create interest.

This human interest in patterns and novelty is a constant push and pull across our lives. We love our regular holiday destination, but every few years we need to mix it up and do something new. We crave stability in relationships, but the positive aspects of stability can turn to negative feelings of boredom and we seek out ways to change things up. We might even have a regular coffee order that one day seems comforting and the next seems constricting, and we get crazy and order a frappuccino instead of a long black. We need same and different all the time and the way we manage this has to do with managing expectations and change.

The ability to manage change and even enjoy it is both a professional and personal trait that countless self-help books for adults deal with. But the embracing of this trait begins very early in childhood and is further molded well into adolescence. This chapter explores how learning music may both assist in developing that skill as well as assisting children who already love change to stick with music learning when it gets frustrating.

Expect the unexpected

Innovation and creativity are buzzwords of the twenty-first century, but without fail when I ask my preservice teachers or an eighth grade class to describe either of those terms, neither group can agree or define it clearly. "I know innovation when I see it" or "Creativity is what creatives do" are two answers that just beg a whole slew of further questions. I'm not going to go down a rabbit hole of possible definitions; however, one thing that is clear regarding innovation and creativity is that they are about breaking the rules. But to break the rules you probably need to know the rules first, and in our current Western education system there are a lot of these to learn and there is a very tricky line between teaching students the rules and teaching students how to think about, question, and effectively break the rules.

Music has a lot of rules, and while this is a whole field of study on its own that I won't go into here, those rules are all about prediction, expectation, and novelty. Music is made up of patterns of sound and we enjoy hearing those patterns repeated or returning throughout a song or piece of music. Just like themes in a story, patterns and themes help a piece of music hang together. We also like the feeling of our expectations being satisfied. Think of a pop song where you know the verse is coming to an end and you are about to return to the familiar and catchy chorus. There might be a complex and pumping drum fill just before the chorus begins, and as the first beat of the chorus comes our expectations are fulfilled and we get a little brain reward rush. Sometimes, after we have heard that satisfying precursor to the chorus six times in a song, the rush isn't there anymore and our brain goes searching for novelty, the new treatment of the old idea. Listen for that moment in a song next time; in the eighties and nineties it was the obligatory key change that helped to keep our brains interested and elongated the song for another minute. In a crime novel or a movie thriller it is the point where you think you know who the baddie is and then the unexpected happens and you say to yourself, "Ooh, interesting."

All of these reactions live in the satisfaction or violation and then resolution of our expectation. As busy modern humans we might favor the former over the latter for the ending of a movie, for instance. We watch a romantic comedy expecting the happy and satisfyingly predictable ending. If we don't get it we are not amused. Ever watched a movie and at the end thought to yourself: Really, is that how you are going to end it? When this happens I often take a moment to think about why it disappointed me—what the filmmakers were trying to say and why this wasn't a satisfying ending.

When studying the impact of personality traits on learning a musical instrument, psychology researchers have named the "Big Five" traits, which are openness, conscientiousness, extroversion, agreeableness, and neuroticism. In a study led by Dr. Kathleen Corrigall, researchers found that two out of the five traits were significantly higher in musicians: conscientiousness and openness to experience. This might be so because "learning to play a musical instrument could be facilitated by *conscientiousness*, which involves self-discipline, organization, and achievement-orientation, and/or by *openness to experience*, which describes the tendency to have an active imagination, to appreciate the arts and literature, to prefer change and variety

over routine, and to be intellectually curious."[7] It can also enhance social interactions via appropriate neural and chemical
activity.

How do these two traits relate to innovation and creativity? A balance of both of these traits is the recipe for creation,
whether that be thinking, making, doing, or viewing differently. The discipline to learn and even master the rules, while
being open enough to know they are in fact only habits and
guidelines and can be changed, can be the basis of innovation.
Even in younger children, violation of the rules with a satisfying resolution fires up the amygdala. The question is, how do
we get from playing flute in a band to having personality traits
that may lead to innovation and creativity?

There is no perfect

If you ever get the chance to sit in on the first few lessons
of a student learning a musical instrument, you will see that
innovation and creativity are not what are encouraged at all.
Actually, it's the opposite. Let's take the flute, for example.
Simultaneously, the student needs to get the right fingers
on the right keys coming from the right direction. In initial
lessons students frequently point the flute out to the left when
it needs to go to the right, put their right hand on the top keys
when it should be their left, and even try to make both hands
come from the bottom over the flute instead of the left hand
from the bottom and the right hand over the top. Confused?
Now you know what it feels like when you start to learn an

---

7   Corrigall, K.A., Schellenberg, E.G. & Misura, N.M. (2013) "Music training,
    cognition, and personality," *Frontiers in Psychology*, 4, p. 222.

instrument and why music teachers use lots of verbal and physical reminders to get the instrument in the right place.

But that's only the beginning. Imagine working really hard to get your flute the right way around and all your limbs in the right place and then you have to make a sound. On the flute, this means putting your bottom lip precisely in the middle of a piece of curved metal and your top lip positioned just right before blowing a column of air into the flute and not passing out from dizziness after losing fifty percent of that air over the top of the mouthpiece. It takes all of a beginner's focused concentration to do everything correctly at the same time—they are working so hard to get these techniques right that there is no room to work outside the rules.

This experience in itself is incredibly beneficial for developing brains. The all-consuming natural act of just getting a note out requires a full brain workout, and when a single note, a short pattern of notes, or an entire piece of music comes out, our brain reward network fires off and we want to do it all over again. But proof that we have succeeded in progressing on our chosen instrument only happens in a given moment and, as I have heard many sportspeople say, you are only as good as your last lap, leap, or shot. Music learning is the same: you are only as good as the last time you played that note, pattern, or piece of music. There is no perfect performance because there is always the next time you get to try to get everything right at the same time.

This phase of learning a musical instrument takes years. To get to the point where a student can put the instrument together without breaking it, put their chair and stand in exactly the right position so they don't poke their neighbor with their bow, and look at the symbol system of music and, without thinking, play the note for the right amount of time and then move on to the next note can be anywhere between two and five years. This is the period of learning the discipline of music, learning the rules or the how-tos. As a parent it is like watching a tree grow: it goes at a glacially slow rate and if you don't pay attention or measure the change in any way, you can easily miss it.

From my experience, high school students accomplish this step in music learning faster than younger children. It is part physical development and heightened control of motor skills, part advanced executive function skills (although this can be hampered by puberty) and part enhanced problem-solving and logic skills. After this stage of music learning is accomplished, students begin to be able to innovate and create, but not in the way you might think. To innovate could be defined in this situation as where we *do* new things, whereas creativity is where we *make* new things. Creating music, or composing it, which is the musically specific term, is easy to identify as a making activity. But innovating—doing something different to improve, problem-solve, or enhance performance—is often both a hidden and very powerful process in music learning.

When students are in a music ensemble situation, say a choir or band, they can't do new things whenever they feel like it. If one person in a sixty-piece band decided to go at a different tempo or speed from everyone else, the performance could well be ruined. But if the trumpet section all miscounted their bars of rest and came in a beat early and the girl on the drum kit heard their mistake and jumped a beat, then her innovation might set the piece back on the rails again. So part of music learning is knowing when to innovate, to do something new, and when to work as a team.

I love sharing a story from my own band that I still can't believe happened. I was conducting a group of mostly senior students in *Barnum*, which is the same story as the movie *The Greatest Showman* as a live musical. We actually did the whole show in a circus tent erected on the school grounds and the band was part of the show, all in costume and visible to the entire audience. It was quite an amazing, high-pressure experience.

One of the songs was performed by a young male character, in this case a ten-year-old boy. As the conductor I would cue him, which means I sort of count him in for the start of his song. They say never work with children or animals because they can and will be unpredictable. Not this young lad—he was on the mark every single time in rehearsal and throughout the shows up until this particular night.

This one night, for no reason at all, he came in a beat early. Just one, but one was enough to throw everything off. In the high-wire act that is live performance, there's no stopping and saying "Let's try that again"—you just have to go on. The audience couldn't see my face as I looked at the band, but the musicians could certainly see the whites of my eyes. What do I do? How do I get the train back on the track when it is still moving?

Most of the time you just need to give in and get to the end of the piece. But here is where the magic happened. The band collectively jumped one beat forward. At the same time. Without discussion or consultation. Twenty-five young musicians collectively innovated and solved the problem in the moment. You couldn't get the smile off my face for the rest of the song!

When we went to the intermission I asked, "Do you know how amazing that was?" True to teenage form, they just shrugged at me and said things like "That was the logical solution" and "What else did you expect us to do?" This is why I love working with teenagers: they have no concept what they are capable of yet and it is my job to help them stretch themselves past what they think they can do.

The subjective or imperfect nature of music learning is sometimes seen as a deficiency, not being able to know if you got it perfectly right and living with the fact that we are human and never *will* get it totally right. But I have seen this imperfection work as a tool to foster the development of innovative thinking skills. There is no completely right, just right for now; there is no one answer, there are many answers. And students get to play with doing new things every time they pick up their instrument.

Seeking out the new

Innovation is doing something new. To do something new we need to know what is the old or the established, and in music learning that is learning how to hold and play our instruments and master the technique to a point where we can break the rules and start exploring. Woven into this is not only the *Where's Waldo?* or spot the odd one out process but also the excitement that comes from seeing the difference and finding the solution, either individually or, as my band did, collectively.

I am always struck by how important learning to value and seek difference is for teenagers. In every school curriculum or set of school values there is always a pointer to the value of diversity, and while I see programs that explicitly teach this disposition, I wonder if we are missing the opportunity to enhance this explicit teaching through the implicit feeling or experience of diversity. So often I see students, and sometimes adults, reacting to a situation where change occurs by instinctively hitting out, and only afterward using their analytical thinking to determine the "right" answer. If we gave more of our students the opportunity to see diversity and change not as good or bad but just different, how might their opinions and reactions change?

Music can provide a tool to help. Music learning is the slow-cook version of valuing and seeking out the new, but music listening can be the instant hit, especially for teenagers. I am not talking about getting your teenagers to love the music you love but instead engaging with them about the music that moves them and why they love it. You may get the teenage rhetoric "you wouldn't understand" but stand strong, become a student of their chosen music or even ask them to make a playlist for you. This opens the door to exchanges on personal likes and dislikes and hopefully a beginning in the valuing of personal choices. While many teachers would say we need to meet the students where they are in their musical tastes, I believe we also have a complementary duty to open their minds to the different, the pattern changes, and the innovations we see. I tell my students, "You don't have to like the music we are studying/playing/listening to but you do have to understand it." Imagine if our almost grown-ups could take that disposition into their next stage of life.

Further reading

Corrigall, K.A., Schellenberg, E.G. & Misura, N.M. (2013) "Music training, cognition, and personality," *Frontiers in Psychology*, 4, p. 222.

Corrigall, K.A. & Schellenberg, E.G. (2015) "Predicting who takes music lessons: Parent and child characteristics," *Frontiers in Psychology*, 6, p. 282.

Tillmann, B. (2012) "Music and language perception: Expectations, structural integration, and cognitive sequencing," *Topics in Cognitive Science*, 4(4), pp. 568–84.

20

CALM THYSELF

Using music to enhance study and manage stress

"Turn that music off! I swear, I have no idea how you can concentrate with that racket."

If you are a parent then you have probably bellowed, or silently seethed, about this at some point. When I present to parent groups this is usually one of the first questions that comes up, followed by an audible sigh, groan, laugh, or eye-roll. The use of music, or sound environments in general, to either promote or hinder productivity is not solely a teenager's domain. Many adults from all walks of life and in wildly different types of work use music, predominantly to be more productive. But the question that comes up most often is this: Is there one genre of music that is better to listen to when studying?

In this chapter we explore the research into music listening and productivity. It is a tricky area to study because there are so many variables: the age of the participants, the type of task they are doing, and whether the study was done under lab conditions or self-reported, to name a few.

Getting into the zone

When I was writing my PhD I needed noise. Not music—noise. I tried different types of music: techno-pop with the same beat all the way through (I got distracted every time the singer came in); inspirational classical music (it activated the classically trained musician inside me and I started figuring out the chord progressions and theme variations); meditation music (made me stare into space before wanting to go to sleep). The noise I needed was not ordered, not immediately musical, and not volume controlled. The noise I needed was the hustle and bustle and whoosh and bang of a barista at work—I needed a cafe.

This fascinated me because I had started to observe behaviors in my students, both teenagers and my preservice teachers, about their music and task choices. The desired outcome was always the same—they wanted to be more productive and complete the task to a high level and quickly—but their music choices to get to that place were based on a few variables: if the task used language, numbers, or graphics; if the task was urgent or progressive; and what type of mood they were in at the time. But there seemed to be another overlay and that was how much music training they had experienced and in what genre of music.

I'm a classically trained musician who loves playing classical music, but I do not listen to a great deal of classical music.

When I look at my PhD writing experience, listening to classical music may have activated my training and pulled my focus away from, rather than toward, the task. As for the techno-pop, maybe it was because I was writing and using language that the song lyrics provoked my brain into multitasking and using the same neural network for language. And meditation music is both slow and very simple musically, which is great for meditation but maybe not so good for the creative spark I needed to write. In a cafe, the noise has a fairly consistent number of sources: coffee machine hissing, banging out and grinding the coffee, quiet talking, and chairs scraping on a hard floor. There is no discernible and distracting structure but a kind of consistency that served to focus my thinking and writing.

Speaking to other doctoral students, however, their choices of music were entirely different, and when I compared it to their music training it became even more interesting. Clive had never touched an instrument in his life and his music of choice for writing was French Europop. He said it was upbeat and kept him feeling positive even when his writing was hard, but it wasn't in a language he understood so the singer just sounded like another instrument. Nellie was a jazz drummer and she couldn't listen to music at all; she preferred to write outside at a cottage that was miles away from anything, saying that nature had no rhythms in it and as soon as she heard a rhythm that was it, pens down on her writing. The more students and adults I spoke to, the more I found that they could clearly articulate what type of music or sound environment made them the most motivated and productive on any given task. This went against the question from parents about the "best" music to listen to and moved toward the idea that we might just have to choose the best music for ourselves.

Listening to music and studying

A study published in 2010 looked at the music-listening habits of students aged from twelve to thirteen, fifteen to sixteen, and twenty to twenty-one years in Japan, Greece, the United Kingdom, and the United States. Six hundred students self-reported in an extensive survey on the types of music they listened to while studying and when, where, and how they chose to do so. One of the significant findings was that "[m]usic played while studying was most strongly reported to relax, alleviate boredom, and help concentration." That might seem a reasonably obvious finding but it takes us back to the parent angst at the start of the chapter and makes me wonder if parents are effectively making their teenager's ability to complete their homework or study more difficult when they give them grief about their music choices.

The study also suggested that students were discerning about their choice of when to play music and when to choose silence: "Students do not play music while studying extensively and that they rarely play music while reviewing for examinations, memorizing material or learning a foreign language and most often play music when thinking or writing. This suggests that they are aware that their performance on some tasks will be impaired, namely those where the cognitive processes involved are shared with those involved in the processing of music." We know that music and language processing have an overlapping neural network so it seems that listening to music with lyrics while writing words is a confusing activity for the brain. While music learning may assist language learning, using music with lyrics when you are trying to write words appears to be counterproductive. Research has found that music without lyrics is more beneficial to our productivity when we are reading or writing, but, to put another twist on it, it only seems to assist thirty percent of participants in the study, while it distracted seventy percent of the participants.

The last quote from this study that I will share is a humdinger for parents and something we probably all have experienced but may not know why it happens: "Across all age groups there was disagreement that they turned music off when someone suggested that they should. This suggests that parents' attempts to prevent music being played while their offspring are studying are likely to be unproductive." Granted that sometimes teenagers in their egocentric fashion turn their music up without regard for the sound environment of others in the house, but this finding makes me wonder if we don't give our teenagers enough credit to make good choices about the sound environment that maximizes their own motivation and productivity. Volume matters just as much as the type of music, too. For me, one loud talker in a cafe and I'm done for (as I wrote these words a loud laugher made herself known, so I may need to move soon). A bigger question for our teenagers might be how they manage their use of music while also instilling in them an understanding of just how important and delicate their hearing is.

Divergent and convergent thinking

Falling under the broader concept of creativity are two types of thinking, divergent and convergent, that are basically different processes that result in creative outputs. I like the definitions that Dr. Simone Ritter and Dr. Sam Ferguson used in their paper looking at which types of music facilitated convergent and divergent thinking: "[d]ivergent thinking involves producing multiple answers from available information by making unexpected combinations, recognizing links among remote associates, or transforming information into unexpected forms," whereas "[c]onvergent thinking emphasizes accuracy and logic, and applies conventional search, recognition, and decision-making strategies."[8] If I had to explain how these different types of thinking felt to teenagers it would be like this: divergent thinking is the quick-fire process of brainstorming where you take off all the restraints and think outside the box, while convergent thinking comes either during or after the divergent brain exposition and looks more closely at which solution might be best and how to put it into action. In business it is blue-sky thinking versus operationalization, but whatever you call it, we use divergent and convergent thinking all the time in order to create.

Ritter and Ferguson's study tested 155 participants to see if there was a type of music that seemed to facilitate divergent thinking, or if silence was better. They compared four types of music (positive, negative, high, and low) that varied in valence (how attractive or "good" the music is) and arousal (how positively our body reacts to the music). Out of the four, classical music had the highest positive valence and high arousal (I am humming Wagner's "Ride of the Valkyries" in my head right now) and facilitated more divergent creative

---

8 Ritter, S.M. & Ferguson, S. (2017) "Happy creativity: Listening to happy music facilitates divergent thinking," *PLOS ONE*, 12(9), e0182210.

thinking compared to silence: Ritter and Ferguson suggest that "the variables involved in the happy music condition may enhance flexibility in thinking, so that additional solutions might be considered by the participant that may not have occurred to them as readily if they were performing the task in silence."

Other research has found that music disrupts verbal working memory as well as tasks where the networks overlap, such as language. From this growing field of research it seems that music enhances some types of thinking and working and disrupts other types, but it is not black or white, good or bad, and there are many variables at play. Again, I wonder if we as parents or teachers either overlay our own personal music enhancement choices on our children or students without taking the type of task into account. In doing so we might just be hindering happiness and cognitive greatness.

Music training and study

So it seems that music listening enhances performance in some tasks and not others and also has a significant effect on a student's state of mind and motivation—sometimes silence is golden and other times it can be a downright killjoy. But does a person's musical training have any impact on this enhancement or otherwise? Through my research I have started to understand that I hear the world entirely differently from the person sitting next to me in a cafe. And what does that fact mean for a class of thirty students or a music ensemble of eighty trained musicians? This question has probably had the most profound effect not only on how I teach but also on how I understand others. In a nutshell, what I have observed about musicians is that they use music as a tool for motivation, creativity, and thinking in a more nuanced and deliberate way.

Originally, many personality or trait-strength tests placed people into one of two boxes: you were either a creative thinker or a logical thinker. This was reinforced by the concept of being left- or right-brained. It was a good place to start while we increased our understanding of the brain and how it functioned and communicated, but in the last fifteen years we have moved far beyond this idea of left *or* right. It is more accurate now to speak about left *and* right, similarly to how we now know it is both nature *and* nurture. As the thinking and research has progressed it has been the study of the musician brain, or some call it the musician advantage, that has helped inform our understanding.

In 2019, Dr. Chia-Chun Wu and Dr. Yi-Nuo Shih completed a study about the use of background noise on work attention performance. Comparing musicians and nonmusicians, they found that when background music was playing the musicians' attention performance was better than the nonmusicians', and that background music tended to improve attention for both groups, but to a greater extent for musicians. Was this because their auditory processing networks were given something to occupy them and that heightened their attention? Could it be that their experience of the sound world is enhanced through their music learning and that stimuli triggers, heightens, or sustains their attention? What I love about this research is that it wasn't published in a music education journal but in the *International Journal of Occupational Safety and Ergonomics*. This sort of research can inform so many more dimensions of our learning and work than just music learning and education.

Much more research is needed in this field but it seems there might be something to my observation that musically trained people interact with the world differently, and possibly in ways that Western societies value and seek to promote.

Sound environments for learning

I would like to add one last observation to this chapter. While delving into the research over the last few years I have continued to teach and have had the great privilege of visiting many schools. I love walking into a new school. For me it is like walking into a science-fiction movie and having to learn all about the new and strange environment. I look at what is on the walls, what the tactile environment is like, even how it smells (I recently walked into a hall of eight hundred teenage boys to give a speech on labels and identity and was struck by the overwhelming smell of boy).

But the sense I use the most acutely, unsurprisingly, is my hearing. What does the school sound like? I'm not talking about volume, though so often we think of sound in a single dimension of loud or soft, and if this chapter has made anything clear I hope it is that less sound is not necessarily better. I can gauge the productiveness, happiness, and connectedness of a school in an instant from how it sounds. Try this sometime when you are waiting to check your bag at an airport: close your eyes for three breaths and listen. Often you'll hear the sound of heightened stress bubbling under the surface.

But back to the school sound environment. I love observing classes, and actually any learning environment gives me a little thrill. I watch the students most of the time, not the teacher, because in the end they are the true measure of how effective a teacher is. It's very common for a teacher to set an individual or small-group task and then, with the best intentions in the world, put on some background music with a view to enhancing the learning.

Here is the thing. With all that we now know about auditory processing, productivity, motivation, individual listening preferences, and music training, is a choice of music made by one person, the teacher, the best idea? I have tested my theory out with teachers and sure enough, teachers who work best in silence favor silent work in their classroom and teachers who need auditory stimulation through music use it regularly in their classrooms. Part of my revelation that everyone hears the world differently would make me question whether my choice of music would work for even a quarter of the class. What then am I doing to the other three quarters of the class through my music choice?

I believe sound environments in schools, workplaces, hospitals, and anywhere else people are productive is a new frontier for development. If children can experiment with different types of music while completing different tasks we could empower them to use music more effectively for their learning and their well-being. Being effective and productive in a personally satisfying way has been identified as a key driver in our human experience of happiness, joy, and purpose. If we update our understanding of how our sound environment impacts these universal aspects of the human condition we could help children, and ourselves, experiment with music as a force for our own good.

Further reading

Gibson, C., Folley, B.S. & Park, S. (2009) "Enhanced divergent thinking and creativity in musicians: A behavioral and near-infrared spectroscopy study," *Brain and Cognition*, 69(1), pp. 162–69.

Kotsopoulou, A. & Hallam, S. (2010) "The perceived impact of playing music while studying: Age and cultural differences," *Educational Studies*, 36(4), pp. 431–40.

Ritter, S.M. & Ferguson, S. (2017) "Happy creativity: Listening to happy music facilitates divergent thinking," *PLOS ONE*, 12(9), e0182210.

Wu, C.C. & Shih, Y.N. (2019) "The effects of background music on the work attention performance between musicians and nonmusicians," *International Journal of Occupational Safety and Ergonomics*, published online February 13, 2019, pp. 1–5.

21

BEYOND SCHOOL

How music learning is the gift that keeps on giving

Everyone has a moment or an interaction in their lives that rocks them to their core. A moment that is so memorable and affecting that you will both never forget it and you will relive in all of its visceral detail every time it comes to mind. One of the most positive moments of my life was as a trainer walking into the Sydney Olympic Stadium with two thousand teenage musicians from all over the world as part of the Sydney 2000 Olympic Marching Band in the opening ceremony. There were 120,000 people clapping and cheering, and when that many people direct the sound they are making in your direction your whole body vibrates. It was like I could feel every particle in my body moving and it was extraordinary.

Another profoundly emotional and this time more negative moment was when the father of a sixteen-year-old student announced to me in a parent-teacher interview that it was time for his son to "put away childish things." He was referring to his son's flute and that, with two years of high school to go, he needed to focus on what was important, meaning science and math and getting a high university entry score.

His son was devastated. He loved playing the flute, though he never wanted to be a professional flautist. He wanted to pursue a career in medicine and while playing flute was fun, that was not the real reason he loved it. The real reason, which he could never articulate to his father in a way that his father would appreciate, was because playing in a band and as part of an orchestra were the only social outlets he had amid the pressure and long hours of studying to attain the results he needed. Sustaining his commitment to achieving this high score across two entire years of school with few opportunities to interact with his peers was very hard and almost his undoing. His well-being and mental health suffered, and while he was no longer allowed to play flute, he sat up in the music department most lunchtimes just to experience some of the culture and companionship he was now missing.

This type of experience begs the question: Should only students committed to becoming a professional musician study music? I like to replace the word music with math in this proposition. Should only students committed to becoming a mathematician study math until the end of high school? The answer to both questions, I believe, is no. Subject area study is in part about the content but in another much larger part about learning skills in one subject area that we can transfer across to others. For example, studying math involves learning how to understand and apply abstract thinking—math as a subject just happens to be doing it using number concepts. Let's take this idea back to the study of music where we could say it involves learning and refining advanced language skills, honing our executive functioning, promoting higher-order thinking, and enhancing social, emotional, and communication skills, all of which can be applied to, well, just about anything. (My apologies to the mathematicians reading this—I am not lessening what students learn in math with a shorter list, I just have more experience abstracting the many nonmusical skills that come from learning music.)

In this final chapter I want to highlight two studies that look at the impacts of music learning after the final school bell has rung. In educational research it is incredibly difficult to separate out the effect of a single activity or influence for a given cohort of students. There are so many variables, such as student age, type of activity, length of activity, teaching approach, and so on. And then there are the social variables, such as parents' education, socioeconomic status, and individual personality. As one of my academic friends used to say regularly, trying to find one single answer to what is the "best" activity for a child is reductionist, which means attempting to take an idea that is very complex and boiling it down to an essential meaning or answer. This can be a useful skill at times, but in our sound-bite, 140-character world we have a tendency to reduce rather than accept the more complex, nuanced, and messy view. Ultimately, we need to look for both viewpoints simultaneously.

After the final bell

In 2015, Dr. Adrian Hille and Dr. Jürgen Schupp from the German Institute for Economic Research in Berlin published a paper looking at the relationship between music learning and skill development. Again, this is research both done outside the field of music education and of broader interest than just what happens in a school day.

Hille and Schupp's comprehensive longitudinal study used a series of data sets, and collected family rather than individual-level data. First, they collected details about the intensity and duration of music activities. Next, data on school results, cognitive skills, personality, time use, and ambition as well as parents' socioeconomic background, personality, involvement in the child's school success, leisure time use as well as taste for the arts were collected. Amassing this broader picture of music learning within the school and home experience helps to see the complex picture and to locate the core or essential findings from all of that data. For those who like numbers, the researchers collected 3,369 samples of students between eight and seventeen years of age. Of that sample group, 372 were musically trained and 2,997 were in the nonmusician control group. Similar data on other activities such as sports were collected as comparison activities.

So enough of the research jargon—what did they discover? Here are some of the most fascinating findings:

- "Even after controlling for a large number of social background characteristics, we find strong differences in terms of cognitive and noncognitive skills between adolescents who learned a musical instrument during childhood and those who did not";

- "Music improves cognitive and noncognitive skills more than twice as much as sports, theater, or dance"; and
- "Our findings suggest that adolescents with music training have better cognitive skills and school grades and are more conscientious, open, and ambitious."[9]

Those wonderful sentences should say it all, but you have stuck with me this far and I can't stop myself. While the second point, which speaks to the effectiveness of music learning in comparison to sports, theater, and dance, could be interpreted as "do music because it is better than anything else," I like to think of it as a very important finding to share. It is probably the more common view for parents that learning music is both a childish pursuit and a reasonably useless one if your child is not a prodigy and planning a career in music. But these findings point to its effectiveness in particular for developing both cognitive and noncognitive skills. Music is an incredible tool for learning as well as a uniquely human art form to enjoy.

In 2019, a population-sized longitudinal study in British Columbia led by Dr. Martin Guhn was published. "Population-sized" means exactly what it sounds like: they looked at a group large enough to be a population, in this case over 112,000 public school students who started school in 2001 to 2003 and followed them through to their final academic high school year. The study, like Hille and Schupp's, controlled for a huge number of variables in order to answer

9   Hille, A. & Schupp, J. (2015) "How learning a musical instrument affects the development of skills," *Economics of Education Review*, 44, pp. 56–82.

questions such as whether there is a significant mean difference between school music participation versus no music
participation with regard to academic learning outcomes in
junior high and high school, and do the patterns of associations
between type of music education and academic achievement
vary by subject area (math, science, English). Two out of the
three researchers were again not from the music education
field but from the School of Population and Public Health
at the University of British Columbia. Their findings don't
look to the health of the music profession but to the health
of the population as a whole, with a focus on government
regulation and support.

Let's get to the findings. The paper is very long and worth
reading, but here are some of the highlights:

- "Participation in school music (especially instrumental music)
  was related to higher exam scores, and students with higher
  levels of school music engagement had higher exam scores.
  The positive relationships between music engagement and
  academic achievement were independent of students' previous
  (Grade 7) achievement, sex, cultural background, and neigh-
  borhood socioeconomic status, and were of considerable
  magnitude";
- "highly engaged instrumental music students were, on
  average, academically over 1 year ahead [in science, math,
  and English] of their peers"; and
- "In light of this study (the largest of its kind to date), as
  well as supporting evidence suggesting music learning in
  childhood may foster competencies (e.g., executive functioning) that support academic achievement, educators may

consider the potential positive influence of school music on students' high school achievement."[10]

These are only two studies, but they are two that draw on and learn from many studies over the previous twenty-five years. They use large populations of participants, follow them for a long time, ask very detailed and specific questions about the type, length, and engagement of the music learning, and control for multiple variables. There are more studies currently in progress and new areas of research emerging, but right now, as I write, we have evidence that should make us stop and consider how we can evolve our educational systems, values and beliefs to let every child have access to the benefits of music learning.

What next?

This is where I should have a list of ten things we can all do right now, wherever we are, but I don't. The reason I don't is because wherever you are reading this, the situation will be different. It will be different because of your school system, the age of your child or your students, the societal values around music and music education, your own personal background and financial status, and probably a thousand other variables.

That doesn't mean you can't do anything, and if you have gotten this far I am hoping there's been at least one "wow" and another "I never knew that" to get you thinking differently

---

10 Guhn, M., Emerson, S.D. & Gouzouasis, P. (2020) "A population-level analysis of associations between school music participation and academic achievement," *Journal of Educational Psychology*, 112, 2, pp. 308–28.

about your own child's or your students' learning journey.
(I hope there's been more but I am managing my expectations.)

Instead of answers, how about I leave you with some obser-
vations and questions to answer in your own context. Almost
all of us are musical from birth, if you look at musicality as the
ability to process music, and we use music right from the beginning of our lives to help understand the world. When we are
very young we use music cues to learn how to communicate
and speak a language. Musical activities, even ones as simple
as clapping our hands, help our brain to connect and synchronize, and something as innate as singing helps us to trust and
feel safe. Music learning before the age of around seven helps
with auditory, visual, and motor cortices' connectivity and
helps the brain's hemispheres and cerebellum communicate
through the corpus callosum. Music learning improves connectivity and symbolization and enhances synaptic speed and
divergent and creative thinking. Music learning enhances our
executive function skills so that we can problem-solve, manage
stress, communicate effectively, and grow into our adult selves.
Music learning allows us to communicate and understand each
other without words and to feel and understand what it is to be
empathic, kind, and productive. It can help human development enormously and, if taught well and effectively, has very
little chance of harming. Why wouldn't we want every child to
have a meaningful and cognitively impactful music education?

So here are the questions for you. What is stopping your
child, your students, your school, your system, your state,
your nation from giving that experience to every child? What
are the beliefs you held about music learning that have been
(hopefully) challenged by what you have read here? How could

you update the understanding your family, school, or community has of how our brains use music as a tool for growth and learning? Where do you sit on the "music for cognitive learning" and "music for its own sake" spectrum and how could both perspectives coexist in your child's life, at least for a little while? How could you move your child's experience of music learning from a fun, nonessential add-on to a daily or weekly progressive learning experience? What is the first question you are going to ask your teenager about their favorite song at the moment? (I left the really hard one until last.)

My goal is very simple: provide a meaningful, ongoing, high-quality music education to every child so that we set them up cognitively to achieve whatever they want to. I tell my preservice teachers they are helping to produce a work of art, and that work of art is a human. They get to put in a few brushstrokes or notes in the melody, but they will never know their full impact, though they may be lucky and see snippets of it one day when a grown-up former student, now holding their own child's hand, comes up and tells them how important they were to them. If I could meet the teacher again who first gave me a clarinet and taught me how to read music, I would probably just cry. And after I had finished blubbering I would say thank you for helping my brain to unscramble itself and learn how to read. You changed my life.

Further reading

Guhn, M., Emerson, S.D. & Gouzouasis, P. (2019) "A population-level analysis of associations between school music participation and academic achievement," *Journal of Educational Psychology*, 112, 2, pp. 308–28.

Hille, A. & Schupp, J. (2015) "How learning a musical instrument affects the development of skills," *Economics of Education Review*, 44, pp. 56–82.

AFTERWORD

In late 2018, the three-part documentary *Don't Stop the Music* aired on Australian national television. It was a simple enough premise: take a school where the students are facing many learning and family challenges, put a music program in place, and see what happens. I remember talking with one of the producers during the filming about how the documentary would be received. He said to me that "it is a really important and very moving story, but it's not a universal one—it is about kids and music, and that just doesn't interest everyone."

I won't say that it interested every Australian but, if viewer numbers, social media engagement, and articles in the media during and for many months after are an indication, it seemed to spark something with far more people than was expected. Maybe the biggest indicator was not just interest and viewers but action. An instrument donation program ran alongside the documentary, and instead of receiving an expected thousand or so instruments through donation, the campaign got more than eight times that number.

I was, of course, so pleased with the success of the documentary and the donation campaign, but I also saw the reactions of people I knew, viewers via Twitter, or even the random commuter in an underground Sydney train station who grabbed me by both shoulders and told me how much the program had moved them. On the outside I listened and shared their excitement and tears about how the documentary had made them think differently about disadvantaged children or music learning, but on the inside I was saying, "But of course music did that."

I kept this feeling to myself; it seemed a bit arrogant coming from the person engaged to be the expert on music education in a documentary about music education. But the feeling kept happening and I wanted to know why.

The answer was that in the way only good documentaries can, *Don't Stop the Music* showed the Australian public something they probably hadn't had the opportunity to see before: the transformative power of the music learning process. So many fictional movies about transformational music programs start with an inspirational teacher, a quirky class of misfits, and an extraordinary effort or belief that culminates in a highly polished, Oscar-worthy performance by the end. This documentary showed the real-life version, although one that was never going to be as polished and perfect. But those involved were so much more than that: they were real people who viewers could identify with, and see themselves or their children reflected in.

The sounds that came out of the instruments were not perfect, far from it. The technique of blowing raspberries on her arm in order to play her trumpet wasn't a textbook response and the sounds coming out the violin were not what one would call pleasant. But as the viewer followed their journey they got to see the little wins, the small successes that are the stuff of music learning transformation. It isn't one big performance that changes lives and brains forever, it is the micro-learnings that happen every time the children pick up their instruments.

I realized a reason I may have been unsurprised by the transformation that everyone else saw was that I get to witness this remarkable change happening several times a year. Visionary and informed school leaders seek my help to implement music programs in their schools, and while the methods and details are always different, the results are invariably similar. Music programs implemented with cognitive and musical benefits in mind, and a sustainability mindset that will see the programs continue long past the exit of the current music teacher or school leader, transform schools, lives, and brains. I specifically mention both cognitive and musical benefits because, for me, it is not an either/or proposition—we don't have music education in the school curriculum only for musical *or* cognitive outcomes. If anything, the past two decades of research have shown that ongoing, sequential music education is necessary for every child to achieve both of these outcomes.

The transformations that I love the most are the unexpected ones. At the very end of the documentary there was a tiny rider explaining that the students in the program had just received their national standardized literacy and numeracy test results and they had, in comparison to previous tests, improved way beyond what was expected. It is hard to explain the improvement without graphs and a two-day workshop in statistics, but it saw them finally achieving at the national standard while in previous tests they were far below it. It is unlikely that any literacy and numeracy program could be as transformative in such a short period of time. All it took was a year of music learning supported by an informed and upskilled teaching staff and an incredible school leader, and maybe a few annoying television cameras hanging around.

My other favorite transformation might seem insignificant, but to me it shows that this music program, along with the others I get to see transforming schools every year, can deliver extraordinary outcomes. Lee Musumeci, the school leader at Challis Community Primary School in *Don't Stop the Music*, was talking to the show's executive producer. Offhandedly at the end of the conversation, after updating her on the brass program's expansion and yet another donation of instruments, she shared that her challenging, huge school, previously with the potential for trouble in all shapes and colors and ready to explode at any moment, was calmer and happier. She quickly clarified that her school was always happy, but that there was a different type of happiness now, one born from a calm place. For a school leader, this part of a school culture can't be bought, trained, or demanded. It has to just happen.

In a calmer school, kindness and care can flourish. In a calmer school, play can happen and friendships can grow. In a calmer school, learning can be consistent, cumulative, and empowering. It seems odd to think that one of the main factors that led to this calmness was often less than perfect sound, but that is the beauty of it. The process of music learning helped these students enhance their cognitive capacity, and while some of them might now hold dreams of becoming professional musicians, that, I believe, is the most pervasive misconception about the place and purpose of music educa-tion in many of our current educational systems.

Music learning can first and foremost help children gain both a love of music and an appreciation of mastery, expertise, and ability. But also, and I believe importantly in our current market economy approach to education, it has the capacity to set the foundations in each child's brain for effective thinking, learning, and understanding. Surely as part of the market economy approach it is clear that meaningful music learning has more to give to every child, not just the talented, interested, and financially able ones, than was previously understood or applied.

Research should both affirm and challenge our current thinking, and in our increasingly evidence-based world we should also ask questions of the research in order to extend the findings. In my 2014 TEDx Talk I asked the question: "What if every child had music education from birth?" I am still asking that question, because while I can't pinpoint exactly what will change or improve, I have seen evidence time and time again of the ability of music learning to propel children, parents, schools, and communities upward in the very areas they have been struggling in. Our children don't just deserve music education—they need it.

ACKNOWLEDGMENTS

When I began working on this book I said to myself, "I've got this, I have written a PhD thesis before, this is the same length and a similar time frame for writing." I had, of course, forgotten how difficult that task was and that while this might have been the second time around, it was still just as hard. For this reason I would like to thank the team at Allen & Unwin for being there for me every step of the way. My deep thanks goes especially to my two publishers: Sandra Rigby, who first approached me and said I think you have a story to tell, and Elizabeth Weiss, who then shepherded this project right to the end. Tom Bailey-Smith and Simone Ford went above and beyond when it came to editing my manuscript and dealing with my mind wanderings.

I would like to thank all of the teachers and students who have let me into their worlds to observe their learning. I get enormous energy from teaching them and honestly, they are my teachers most of the time. In particular I would like to thank principals Lisa Parrello and Lee Musumeci for inviting me to work with their staff and most of all with their students. I am also very grateful to Paul Scott-Williams and Keva Abotomey for believing in and supporting me to make the

Goulburn Strings Project a reality and to Tara Smith and Vicki Norton for involving me in the Australian Chamber Orchestra Foundations Program. I would also like to thank the incredible staff of Canberra Grammar School, past and present, who have supported me always to reach further than I think I can. My particular thanks to Dr. Justin Garrick, who continues to inspire me with his leadership and vision.

I would also like to thank ABC and Artemis Media for giving me the opportunity to share the experience of music learning with the Australian public through the *Don't Stop the Music* documentary. It did change just about everything, as you said it would. In particular I thank Celia Tate, Steve Westh, Teri Calder, Stephen Oliver, Aden Date, and, of course, Guy Sebastian, for guiding me through that journey.

I greatly appreciate the support of Professor Alan Harvey and Dr. Kat McFerrin as reviewers for this book. I have suffered more than one bout of impostor syndrome during the writing of it but your expertise and gentle support when I missed the mark was invaluable to me.

I have a team of very patient people who keep the wheels of my work turning when I need to hunker down and write. My deepest thanks to them for driving the ship when the captain went AWOL.

Lastly to my family, who have supported me through this journey and the writing of this book. Mum, you were my first inspiration as a teacher and the educator I look up to every day. Ellie, you are the reason why I am trying to improve the lives of every child, and, Michael, you were there to support me when I wrote 500 good words or 1,000 bad ones.

Thank you.

INDEX

Dr. Anita Collins is an award-winning educator, researcher, and writer in the field of brain development and music learning. She is internationally recognized for her unique work in translating the scientific research of neuroscientists and psychologists for parents, teachers, and students. In 2014, Anita wrote one of the most watched TED Education films ever made, *How Playing an Instrument Benefits Your Brain*, which led to an invitation to speak at TEDxCanberra later that year. Anita regularly presents her research on television, radio, and through her scholarly and popular writings. She was the lead on-screen expert in the ABC-TV series *Don't Stop the Music* and is a music teacher and conductor at Canberra Grammar School, assistant professor in music and arts education at the University of Canberra, and associate fellow of music, mind and wellbeing at the University of Melbourne.